T0134822

SxI - Springer for Innovation /
SxI - Springer per l'Innovazione

Volume 15

More information about this series at http://www.springer.com/series/10062

Peregrina Quintela Estévez •
Bartomeu Coll • Rosa M. Crujeiras •
José Durany • Laureano Escudero
Editors

Advances on Links Between Mathematics and Industry

CTMI 2019

Editors
Peregrina Quintela Estévez
ITMATI (Technological Institute for Industrial Mathematics), R&D&i Unit
Santiago de Compostela, Spain

Department of Applied Mathematics
Universidade de Santiago de Compostela
Santiago de Compostela, Spain

Red Española Matemática-Industria (math-in)
Santiago de Compostela, Spain

Rosa M. Crujeiras
ITMATI (Technological Institute for Industrial Mathematics), R&D&i Unit
Santiago de Compostela, Spain

Department of Statistics, Mathematical Analysis and Optimization, Universidade de Santiago de Compostela
Santiago de Compostela, Spain

Laureano Escudero
Red Española Matemática Industria (Math-in)
Santiago de Compostela, Spain

Area of Statistics and Operations Research
Universidad Rey Juan Carlos
Madrid, Spain

Bartomeu Coll
Department of Mathematics and Computer Science, Universitat de les Illes Balears (UIB)
Palma, Spain

Red Española Matemática – Industria (math-in)
Santiago de Compostela, Spain

José Durany
ITMATI (Technological Institute for Industrial Mathematics), R&D&i Unit
Santiago de Compostela, Spain

Department of Applied Mathematics
Universidade de Vigo
Vigo, Spain

ISSN 2239-2688 ISSN 2239-2696 (electronic)
SxI - Springer for Innovation / SxI - Springer per l'Innovazione
ISBN 978-3-030-59222-6 ISBN 978-3-030-59223-3 (eBook)
https://doi.org/10.1007/978-3-030-59223-3

This Springer imprint is published by the registered company Springer Nature Switzerland AG
The registered company address is: Gewerbestrasse 11, 6330 Cham, Switzerland

Preface

These Proceedings collect a set of selected works that have been presented at the *First Conference on Transfer between Mathematics & Industry*, CTMI 2019, held at Santiago de Compostela, Galicia, Spain, July 22–24, 2019. It is the first edition of a conference exclusively dedicated to the transfer between mathematics and industry. The objective is to turn it into a biannual forum that focuses its activity on three huge objectives: facilitate the industrial innovation, improve the connection between Academy and Industry, and convey to the society the role of mathematics in the world that surrounds us.

The conference has been promoted by the *Spanish Network for Mathematics & Industry* (math-in) and it was organized in collaboration with its partner, the *Technological Institute for Industrial Mathematics* (ITMATI).

The math-in network is a nonprofit association, launched in 2011, formed by 30+ mathematical research groups that belong to Spanish universities and research centers. The mission of math-in consists of providing solutions and mathematical technology transferring to productive sectors of the Society, included by industrial enterprises and the public administrations.

ITMATI is a public consortium of the three Galician universities devoted to provide advanced solutions for productive sectors of society, especially businesses, industries, and the public administration, helping to improve competitiveness and supporting innovation in the productive sector.

Mathematics is a strategic good that accelerates economic growth, and affects all sectors, in the different phases of the productive process: from the design, modeling, simulation and prototyping of products, to the optimization of productive and organizational processes and the analysis of data. Both institutions, math-in and ITMATI, were born with the aim to booster mathematical impact on the innovation, visualizing them as key technologies to all economic sectors.

The CTMI 2019 structure included four invited speakers, five thematic mini-symposia, two communication sessions, and two sessions of success stories posters. In addition, during the conference, a matchmaking event Mathematics and Industry was organized by the *Enterprise Europe Network*. CTMI 2019 brought together leading international personalities from business, science, and academia. This book of proceedings tries to give an overview of all its program through its eight chapters to highlight several real-world applications of industrial mathematics:

- Chapter “Recent Advances in Computational Models for the Discrete and Continuous Optimization of Industrial Process Systems”, by Hector D. Perez and Ignacio E. Grossmann, presents a comprehensive overview of the main disciplines of mathematical optimization (linear and nonlinear, and deterministic and considering uncertainty with several variants of stochastic and robust optimization, among others), considering continuous, binary and integer variables and a mixture of them. Some decomposition algorithms for problem-solving are dealt with. Up to 20+ different projects that are applied to industry are also considered, for different end-user companies. They have been developed by teams, whose PI has been Prof. Ignacio Grossmann at the Center for Advanced Process Decision-Making (CAPD), Carnegie Mellon University, Pittsburg, USA. The application fields that are considered are Oil and Gas upstream operations, material blending facilities, natural gas plant design, biofuels synthesis, supply chain facility network design, water network design, industrial electricity market integration, reliable plant design, and supply chain design.
- Chapter “Optimal Design of a Railway Bypass at Parga, Northwest of Spain”, by Gerardo Casal, Alberte Castro, Duarte Santamarina, and Miguel E. Vázquez-Méndez, collects different models that have been proposed during the past years to obtain the optimal design of a linear infrastructure (road or railway) connecting two given points. The usefulness of one of these models to design a railway bypass is analyzed by considering a real case study on the railway line A Coruña-Palencia (Spain), where it passes through the urban area of Parga. The good performance of the model is reported by using an ad hoc algorithm for problem-solving.
- Chapter “Reduced Models for Liquid Food Packaging Systems”, by Nicola Parolini, Chiara Riccobene and Elisa Schenone, revisits a set of reduced numerical models that have been developed in the past few years. They support the design of paperboard packaging systems. Currently, simulation tools are elements for effective production, design, and maintenance processes in various industrial applications, accounting for the complete three-dimensional complex physics. However, it is argued that these tools are not always the best option to pursue, in the preliminary design phase or whenever very fast evaluations are required.
- Chapter “Reduced-Order Modeling in the Manufacturing Process of Wire Rod: Applications for Fast Temperature Predictions and Optimal Selection of Process Parameters”, by Elena B. Martín, Fernando Varas and Iván Viéitez, considers that the high number of operational variables that determine the cooling process of steel wire, given by the conveyor velocity and the different fan sections powers, lead to a dependency of the cooling on a high multidimensional parameter space. Given the potentially high number of combinations, an efficient strategy is presented, based on the use of Higher Order Singular Value Decomposition. The approach provides a Reduced Order Model (ROM). It can predict quite accurately the cooling curve for any combination of the process parameters. Also, the ROM in combination with an optimization tool finds the

adequate operational parameters with a significant reduction of energy consumption.

- Chapter "Modeling and Numerical Simulation of the Quenching Heat Treatment. Application to the Industrial Quenching of Automotive Spindles", by Carlos Coroas and Elena B. Martín, studies the quenching heat treatment. It consists of the immersion of a steel piece in a fluid. The fast cooling undergone by the piece induces microstructure transformations aimed to provide the piece with specific mechanical. The numerical model needed to mimic the cooling process is studied in order to predict the crystallographic structure. The final model is eventually used to optimize the manufacturing parameters of the steel industrial quenching process of spindles in the automotive industry. As a second part of the work, various components of the Wildfire Resources Management (WRM) are computationally studied, by using solvers such as SYMPHONY and Gurobi.
- Chapter "Single Particle Models for the Numerical Simulation of Lithium-Ion Cells", by Alfredo Ríos-Alborés and Jerónimo Rodríguez, considers that in the design of Battery Management Systems (BMS) for a lithium-ion cell, it is crucial to accurately simulate the device in real time using mathematical models. Often, Equivalent Circuit Models (ECM) are used to this end, due to their simplicity and efficiency. However, they are purely phenomenological, and their internal variables lack physical meaning. In this work, the Single Particle Model (SPM) is reviewed. It is a physics-based model of reduced complexity that is suitable for real-time applications.
- Chapter "Fracture Propagation Using a Phase Field Approach", by David Casasnovas and Ángel Rivero, provides a brief introduction on the theory of phase field models. They have received a lot of attention during the last 20 years. A review is also presented on some striking applications to show the huge potential and versatility of the technique. In particular, a simple model is presented to study the propagation of fractures in elastic homogeneous materials. It eventually evolves into a coupled model that includes fracture propagation and flow in elastic-porous media.
- Chapter "Phase Space Learning with Neural Networks", by Jaime López García and Ángel Rivero, proposes an autoencoder neural network as a nonlinear generalization of projection-based methods for solving Partial Differential Equations (PDEs). The proposed deep learning architecture can generate the dynamics of PDEs, by integrating them completely in a very reduced latent space without intermediate reconstructions. The learned latent trajectories are represented, and their physical plausibility is analyzed. It is shown the reliability of properly regularized neural networks to learn the global characteristics of a dynamical system's phase space.

We would like to address our warmest thanks and gratitude to all who have made this book possible. First of all, to all the speakers of CTMI 2019 for their valuable contributions and, very specially, to those who accepted our invitation to contribute to this proceedings volume. Next, to the anonymous referees that helped the authors to improve the quality of their manuscripts. We also would like to thank the support of the *European Service Network of Mathematics for Industry and Innovation* (EU-MATHS-IN), the *Galician Innovation Agency of the Xunta de Galicia* (GAIN), the Spanish *Mathematics Strategic Network* (REM), the *Enterprise Europe Network*, and the *Faculty of Mathematics of the Universidade de Santiago de Compostela*. Many thanks to the members of the Scientific Committee of CTMI 2019, for their wise choice of the central topics of the conference and the invited speakers; their work has been essential to the success of the conference. We also thank the Springer staff for their help and support during the edition process, and very specially Francesca Bonadei. Thanks also to Adriana Castro of ITMATI for her important work coordinating the publication of this book.

We, members of the Organizing Committee and also editors of this volume, were deeply involved in the preparation of CTMI 2019 and very much enjoyed this experience especially in working with the numerous participants. We are happy to complete our work by editing this volume and wish that you will find the CTMI 2019 proceedings interesting, stimulating, and a great experience.

Santiago de Compostela, Spain
July 2020

Peregrina Quintela Estévez
Bartomeu Coll
Rosa M. Crujeiras
José Durany
Laureano Escudero

Contents

Recent Advances in Computational Models for the Discrete and Continuous Optimization of Industrial Process Systems

Hector D. Perez and Ignacio E. Grossmann

Abstract

An overview of the mathematical formulations used for discrete and continuous optimization are presented. These include Linear Programming, Nonlinear Programming, Integer Programming, Mixed-Integer Linear Programming, Mixed-Integer Nonlinear Programming, Logic-based Optimization, Stochastic Programming, Robust Optimization, and Flexibility Analysis. Successful applications of optimization models in industry are presented in the following fields: upstream oil & gas, materials blending, natural gas, biofuels, water treatment, electricity market integration, plant reliability, and supply chain design. Ongoing projects applying computational models to optimize industrial process systems are also mentioned. Implementations of customized optimization techniques that improve computational performance and enable finding solutions to otherwise unsolvable optimization problems are highlighted. These include strengthening cuts, decomposition strategies, model reformulation, and linearization, among others.

1 Introduction

Continuous and discrete optimization has played an important role in improving industrial processes. Since its origins, the field of optimization has been influenced by key researchers from the process industries. Among these are Martin Beale, Jacques F. Benders, Abraham Charnes, and William W. Cooper, who were key figures for applying mathematical programming in the oil industry. Beale joined the

H. D. Perez · I. E. Grossmann (✉)
Department of Chemical Engineering, Carnegie Mellon University, Pittsburgh, PA 15213, USA
e-mail: grossmann@cmu.edu

P. Quintela Estévez et al. (eds.), *Advances on Links Between Mathematics and Industry*, SxI - Springer for Innovation / SxI - Springer per l'Innovazione 15,
https://doi.org/10.1007/978-3-030-59223-3_1

Corporation for Economic and Industrial Research (CEIR), in 1961, which later became Scicon (Scientific Control Systems Ltd.), where he led the development of mathematical programming software for industrial applications [3, 42]. Benders joined the Shell laboratory in Amsterdam, in 1955, and applied mathematical programming techniques to oil refinery logistics [4]. Charnes and Cooper were both affiliated with what is currently Carnegie Mellon University. Their research included applications of mathematical programming for aviation fuel blending in collaboration with Gulf Oil [14, 15].

The evolution of applied industrial optimization has led to the birth of a field called Enterprise-wide Optimization (EWO) [22]. EWO targets a more complete view of industrial processes that includes not only manufacturing, but also supply and distribution within the enterprise. There are different layers to EWO, namely, planning, scheduling, and control. These are depicted in the decision-making pyramid shown in Fig. 1. The distinguishing element between the three decision levels is the time scale of the events involved. The control level pertains to the second to minute operational decisions at the manufacturing facilities, involving the manipulation of equipment and process parameters. The scheduling level involves decisions at the hours and days resolution such as the allocation of resources and event sequencing. The planning level is for long-term decisions (weeks to years resolution) such as long-term investment decisions and operational targets. Within each level, optimization targets include profit maximization, improved resource utilization, cost minimization, and sustainable design and operation. Optimization in each of these decision levels can bring significant benefits to industries where implemented.

The content of this paper is organized in the following sections. Section 2 gives an overview of research focus areas in EWO at Carnegie Mellon University, highlighting industrial collaborations. Section 3 presents an overview of the different types of continuous and discrete optimization models used for EWO. Section 4 presents a total of 8 industrial areas where optimization has been applied. Concluding remarks are given in Sect. 5.

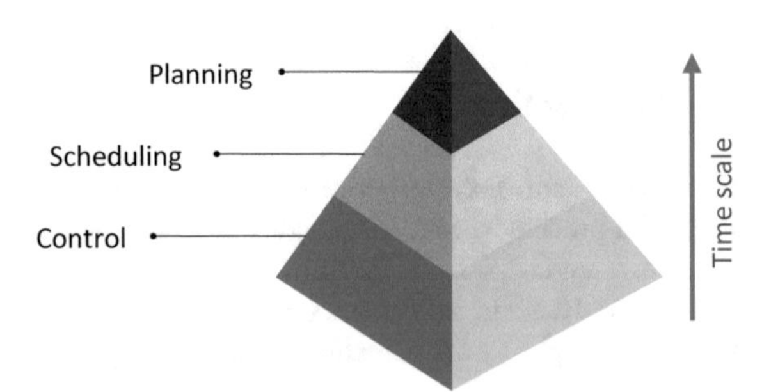

Fig. 1 Decision-making pyramid for EWO

2 EWO Group

At Carnegie Mellon University, a multidisciplinary research center now known as the Center for Advanced Process Decision-making (CAPD) was established in the mid-80s, by professors Art Westerberg, Larry Biegler, and Ignacio Grossmann. This center has grown since and is composed of researchers in the fields of chemical engineering, operations research, and industrial engineering. Its research goals include: (1) understanding and supporting complex issues faced by industry both from a design and an operational point of view, and (2) developing modeling and solution techniques to address those issues. Within the CAPD, there is an interest group that focusses on Enterprise-wide Optimization (EWO), whose goals include (1) optimizing entire supply chains, (2) developing novel planning and scheduling models that oftentimes address uncertainty, and (3) integrating planning, scheduling, and real-time optimization. A wide portfolio of projects has been undertaken over the years by the EWO group. Recent projects in the group along with the company collaborators are listed below.

1. Network design, planning, and scheduling
    a. Demand side management in the steel industry (*ABB*)
    b. New copper concentrate optimal scheduling topology (*Aurubis*)
    c. Unconventional oil gathering network design (*ExxonMobil*)
    d. Stochastic multi-period oilfield planning and design (*SKInnovation*)
    e. Integration of reservoir modeling with oilfield planning (*Total*)
2. Reaction systems
    a. Residue fluidized catalytic cracking real-time optimization (*Petrobras*)
    b. Continuous reactor design and optimization (*Dow*)
    c. Kinetic parameter estimation (*Eli-Lilly*)
3. Algorithms
    a. Stochastic non-convex mixed-integer nonlinear programming algorithms (*ExxonMobil*)
    b. Design space via symbolic computation (*Eli-Lilly*)
4. Data-driven approaches
    a. Polymer design with derivative-free optimization (*ExxonMobil*)
    b. Data-driven optimization of integrated chemical plants (*Dow*)
    c. Production schedule optimization with deep reinforcement learning (*Dow*)

5. Equipment and Maintenance

    a. Integrated reliability and storage design with maintenance policy optimization (*Linde*)
    b. Heat exchanger circuitry optimization (*MERL*)
    c. Advanced heat exchanger model optimization (*UTI Carrier*)

6. Supply Chain and Logistics

    a. Novel continuous-time inventory routing algorithms (*Air Liquide*)
    b. Multi-period vehicle routing algorithms (*Linde*)
    c. Full truckload delivery planning (*Braskem*)
    d. Customer service marginal cost estimation (*Air Liquide*)
    e. Digital supply chain business process optimization (*Dow*)
    f. Portfolio-wide optimization in the pharmaceutical industry (*Eli-Lilly*).

Industrial collaboration has been key to the research success and knowledge development in the EWO group. Industry driven projects have provided key insights that give real world relevance to the projects developed in the group. The knowledge gained from these collaborations is freely available to the public at http://egon.cheme.cmu.edu/ewo/seminars.html. In regard to the projects undertaken by the EWO group, most projects apply methods from mathematical programming, with increased emphasis on uncertainty. There has also been interest in other optimization approaches such as data-driven modeling, artificial intelligence, and symbolic computation.

The following section will provide an overview of the mathematical programming models used in discrete and continuous optimization.

3 Optimization Models

The general formulation for continuous and discrete optimization consists of an objective function and a set of constraints as shown in (1). The objective function f can be a linear or nonlinear single-valued function. The optimization sense is usually minimization, although maximization can also be used. The set of constraints is given by linear or nonlinear sets of constraint inequalities and equalities. In the general notation given below, g_i is a vector-valued function of the variables x and y, which represent continuous and discrete variables, respectively. Other continuous and discrete sets besides m-dimensional reals and n-dimensional integers can be used for the domains of x and y. Binary variables are often used for the discrete variables to denote yes or no decisions.

$$\begin{aligned} &\min_{x,y} f(x,y) \\ &s.t. \quad g_1(x,y) \le 0 \\ &\qquad g_2(x,y) = 0 \\ &\qquad g_3(x,y) \ge 0 \\ &\qquad x \in \mathbb{R}^m \\ &\qquad y \in \mathbb{Z}^n \end{aligned} \tag{1}$$

Mathematical formulations can be divided into the following groups:

- Linear Programing (LP): all expressions are linear with respect to the variables, and only continuous variables are used. Algorithms for solving LPs fall into two categories: pivoting methods (e.g., simplex) and barrier methods (interior-point methods) [29, 46].
- Nonlinear Programing (NLP): at least one nonlinear expression is used, and only continuous variables are used. Algorithms for solving NLPs fall into three main categories: (1) reduced gradient methods (e.g., GRG2, CONOPT, and MINOS codes), (2) successive quadratic programming (e.g., SQP code), and (3) interior-point methods (e.g., IPOPT code) [8].
- Integer Programming (IP): all expressions are linear, and all variables are integer-valued. The two main algorithms for solving IPs are branch-and-bound [16], cutting planes [2], and branch-and-cut [31]. Many integer programs fall under the category of 0-1 IP, where all variables are binary.
- Mixed-Integer Programming (MIP): both continuous and discrete variables are used. When all expressions are linear, the framework is referred to as mixed-integer linear programming (MILP). When at least one expression is nonlinear, it is referred to as mixed-integer nonlinear programming (MINLP). Algorithms for MILPs are the same ones as for IPs. In the case of MINLPs, the main algorithms are branch-and-bound, Generalized Benders Decomposition (GBD) [21], Outer-Approximation methods (OA and QOA) [18, 45], and the Extended Cutting Plane method (ECP) [48]. MINLPs can at times be linearized via exact linearization or piecewise linear approximations to improve their solvability.

Logic-based formulations have also been developed under the framework of Generalized Disjunctive Programming (GDP) [43]. The GDP formulation is given in (2), where c_k is a scalar fixed cost for disjunction k, f is a single-valued function, $g(x) \le 0$ represents the set of common constraints among scenarios. The OR operator ($\vee$) is used to select among a set of alternatives J_k in the set of disjunctions K. When a given alternative is selected, the respective Boolean variable Y is *True* and activates the constraints of the form $r_{j,k}(x) \le 0$ and sets the fixed cost to $\gamma_{j,k}$. The vector-valued Boolean function $\Omega(Y)$ is used to incorporate additional logical expressions, which must be *True* for the solution to be feasible.

$$
\begin{aligned}
\min_{x,Y} \; & \sum_k c_k + f(x) \\
s.t. \; & g(x) \leq 0 \\
& \bigvee_{j \in J_k} \begin{bmatrix} Y_{j,k} \\ r_{j,k}(x) \leq 0 \\ c_k = \gamma_{j,k} \end{bmatrix} \forall k \in K \\
& \Omega(Y) = True \\
& x \in \mathbb{R}^m \\
& Y_{j,k} \in \{True, False\}
\end{aligned}
\tag{2}
$$

The GDP formulation can be reformulated into 0-1 MILP/MINLP problems. In the process, Boolean variables are translated into binary variables. The two main reformulation techniques are the Big-M method and the Hull Reformulation method [25]. There is a tradeoff between the two in terms of model complexity and model tightness, with the latter resulting in tighter, but more complex models.

Constraint programming (CP) [28] is a logic-based framework that can be applied to optimization. It is well suited for scheduling problems, in which resources are limited and relies on an efficient method for finding feasible solutions via a technique called constraint propagation. Hybrid CP/MILP approaches have also been used to combine the advantages of both methods [30, 38]. For instance, CP can be used for the sequencing constraints and MILP can be used for the resource assignment constraints. This technique has promising synergy that can significantly improve solution times.

Optimizing under uncertainty plays an important role in industrial applications, in which many different types of uncertainties can be encountered [44]. Uncertainties can be classified as exogenous or endogenous, depending on what triggers the realization or disclosure of the uncertain parameter values [32]. When the triggering is external to the decision maker (e.g., future demand for a product), the uncertainty is termed exogenous, and when it is triggered by the decision maker's choices (e.g., oil field size, which can only be known after drilling and producing), it is termed endogenous. The three main approaches to optimization under uncertainty are Stochastic Programming [9], Robust Optimization [6], and Flexibility Analysis [24].

The most common stochastic programming (SP) approach is the two-stage formulation, which is presented in (3). The objective function in stochastic programming is an expectation of f given the uncertain parameter ξ, which has a discretized probability distribution. The two-stage formulation separates the here-and-now decisions (x; stage 1) from the wait-and-see decisions (y; stage 2). The latter are recourse decisions that are made in response to a scenario s. A scenario is a state that arises from one of the possible realizations of the uncertain parameters. The formulation can be extended to multistage scenarios. Multistage scenarios are often represented with a scenario tree, as the one shown in Fig. 2, for a three-stage stochastic program under exogenous uncertainty with two realizations

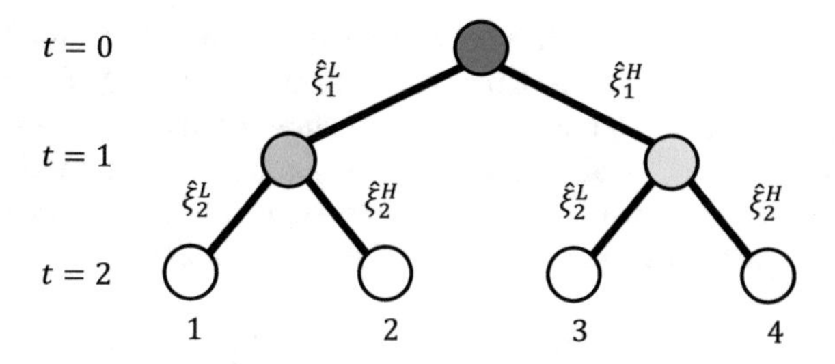

Fig. 2 Sample scenario tree for a three-stage stochastic program with exogenous uncertainty

(high and low) for the uncertain parameter ξ in each stage. Nodes represent the states of the system at each period and arcs represent transitions from one period to the next.

$$\begin{aligned} \min_{x,y} \quad & \sum_{s \in S} p_s \cdot f(x, y_s, \xi_s) \\ s.t. \quad & g(x, y, \xi) \leq 0 \\ & x \in X \\ & y \in Y \end{aligned} \tag{3}$$

Two important metrics are typically associated with the use of stochastic programming: EVPI (expected value of perfect information) and VSS (value of the stochastic solution). The EVPI quantifies the improvement that is possible if uncertainties are replaced with perfect information. This gives an indication of how valuable it would be to reduce uncertainties by improving forecasting accuracy, for instance. On the other hand, the VSS quantifies the added value of using stochastic programming relative to the deterministic optimization counterpart. These metrics are typically applied to two-stage problems, but can be extended to multistage problems as discussed in the work by Escudero, et al. [19].

Robust optimization (RO) takes a different approach to uncertainty. Rather than optimizing over possible scenarios given uncertain parameter probability distributions, the focus is on optimizing over a deterministic uncertainty set to guarantee feasibility over all points in the set. The problems can be formulated so that the degree of conservatism is customizable [7]. The general formulation for RO is given in (4). This formulation is aimed at finding fixed operating variables x that minimize the maximal value of the constraint system under the uncertainty set W. This ensures the solution is feasible for all uncertain points since it is feasible for the worst-case uncertainties in W.

$$\min_{x \in X} \max_{w \in W} \max_{j \in J} f_j(x, w) \tag{4}$$

Extensions to the classical RO approach are Adjustable Robust Optimization (ARO) and Affinely Adjustable Robust Optimization (AARO) [5]. In these formulations, the operating variables are a function of the uncertain parameters, meaning $x = x(w)$. In AARO, the relationship between the uncertain parameters and the operating variables is linear, $x = p + Qw$. These approaches have the advantage of increased flexibility in the formulations for more realistic optimization.

Another approach to optimizing over an uncertain set is that of Flexibility Analysis (FA) [23, 24], which uses an alternate sequence of optimization operators than that of RO. Error! Reference source not found. provides the general formulation for FA, which finds the worst parameter w by maximizing over the uncertainty set W, given that the operating variables x are adjusted for each parameter w to minimize the largest constraint in the system. For linear models, FA problems are translated into MILPs, whereas AARO problems are translated into LPs. Although FA is computationally more expensive given its more general treatment of the operating variables, requiring the use of binary variables in its reformulation as an MILP, it provides solutions that are feasible and more rigorous in the uncertainty set when compared to AARO [52]. Thus, FA can find better solutions than those proposed by AARO.

$$\max_{w \in W} \min_{x \in X} \max_{j \in J} f_j(x, w) \tag{5}$$

4 Industrial Examples

4.1 Upstream Oil and Gas

According to a market research report by IBISWorld [41], the Oil & Gas E&P (Exploration and Production) industry generated \$3.3 trillion USD in revenue in 2019. Of that revenue, roughly 75% came from oil and 25% from gas. The industry as a whole is very capital intensive with a 4:1 capital investment to labor costs ratio. CAPEX (capital expense) for production facilities can be on the order of billions of dollars in the case of offshore facilities. Optimization in this industry can generate significant value, as even small percentages in cost reduction or revenue increase can amount to millions of dollars. As a result, several studies have focused on the optimization of both offshore and onshore facilities. Industrial optimization examples in each of these areas are presented in this section.

4.1.1 Offshore Deep-Water Oilfield Development

An application of mathematical programming to plan the optimal development of deep-water oil fields is presented in the work by Gupta and Grossmann [26]. A multi-period non-convex MINLP model is used to plan the development and production of multiple oil fields. The model accounts for simplified reservoir models and multicomponent (oil, gas, and water) systems. Key decisions of the offshore development planning model belong to the following areas:

- Installation and expansion timing for FPSO (floating production and storage) units,
- FPSO unit capacities,
- FPSO connections to the oil fields,
- number of wells to drill in each field,
- oil and gas production rates at each field.

The objective of the model is to maximize the NPV (net present value) of the offshore development project. High order polynomials for the reservoir profiles in terms of cumulative water and cumulative gas production are used to avoid the bilinear terms resulting from using water-oil ratios or gas-oil ratios. The model can also be reformulated as an MILP by using exact and piecewise linearization techniques to find the global optimum of the approximated problem. Figure 3 illustrates a piecewise linearization of a nonlinear oil delivery profile.

A case study for the above model is presented for a project with a 20-year horizon, 10 oil fields, 3 FPSOs, 23 wells, 3-year lead time for FPSO installation, and 1-year lead time for facility expansion. The MINLP model has approximately 500 binary variables, 5,700 continuous variables, and 9,900 constraints. It is solved in 67 s using DICOPT 2x-C on an Intel Core i7 with 4 GB of RAM to yield an optimal NPV of $30.95 billion USD. The reformulated MILP has approximately 1,600 binary variables, 12,000 continuous variables, and 17,000 constraints and requires a considerably longer computational time (approx. 4.5 h) to find a global optimum of $30.99 billion USD. Connection schedules and production rates are shown in Figs. 4 and 5. The three FPSOs are installed within the first two years and production begins in the end of year three (beginning of year four), with additional field connections scheduled for the following three years.

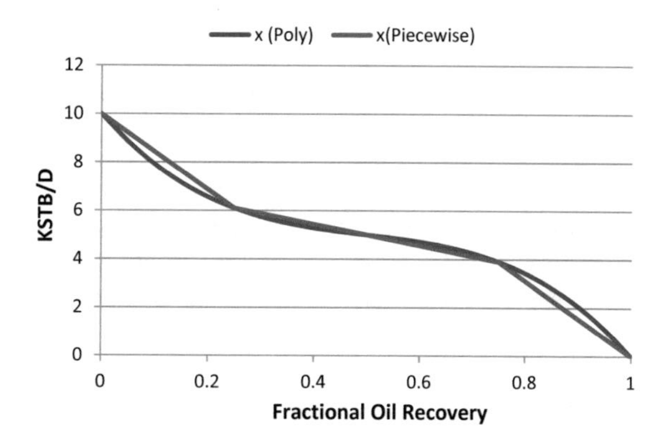

Fig. 3 Sample piecewise linearization for oil deliverability profiles

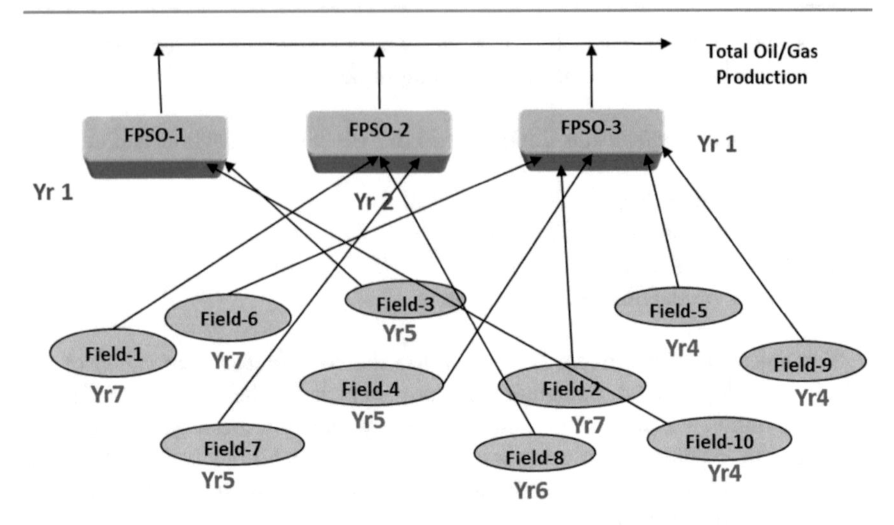

Fig. 4 Installation plan for deep-water case study (Adapted with permission from [26]. Copyright (2012) American Chemical Society)

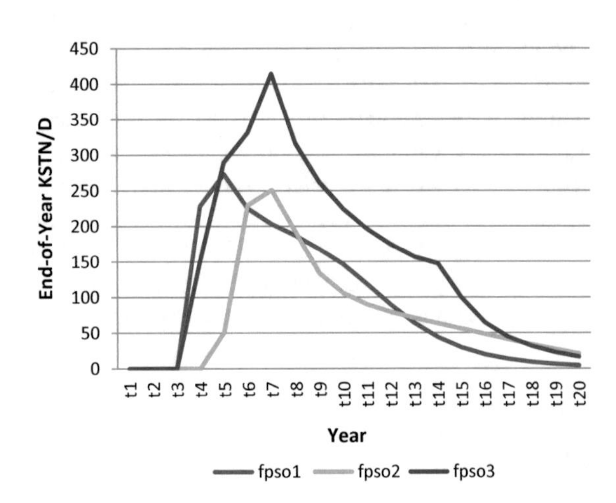

Fig. 5 Oil production rates for deep-water case study (Adapted with permission from [26]. Copyright (2012) American Chemical Society)

The 2012 deterministic model discussed above has been extended to a multistage stochastic model that accounts for (1) endogenous uncertainties in the oil field parameters and (2) fiscal contracts involving taxation and royalties [27]. The revenue distribution included in the model is as follows,

1. Contractor's share

   a. Oil cost
   b. Share from After-tax profit

2. Government's share

   a. Royalties
   b. Direct share from profit
   c. Indirect share from profit (income tax on contractor).

Langrangean decomposition is used to parallelize and improve the computational performance during solution time. For a project with 3 fields and three FPSOs, studies show the benefits of using decomposition techniques for solving stochastic programs. When 4 scenarios involving uncertainties in the oil field size, oil deliverability, water-oil ratio, and gas-oil ratio are included in the project, the full-space model is solved to optimality using CPLEX 12.2 in under 3 h, yielding an expected NPV of $12 billion. Decomposition yields a solution that is only 0.5% below the optimum and has an optimality gap of less than 2%. However, the solution times for the decomposed model are 7.3 and 4.3 min for the sequential and parallel implementations, respectively. This is a significant speedup of 2 orders of magnitude.

Another instance that includes uncertainty in the oil field size (4 scenarios) and accounts for progressive production sharing agreements shows that the sequential and parallel decomposition implementations find a solution of $3 billion with an optimality gap of 0.7% in only 2 h and 1 h, respectively. On the other hand, the full-space model times out after 10 h with a 21% optimality gap and a solution that is 2.3% below the solution obtained via decomposition. Larger instances with 5 oil fields and up to 8 scenarios show that even though decomposition strategies become more computationally expensive, they find better solutions with tighter bounds much faster than the full-space models.

4.1.2 Onshore Shale Gas Development

A multi-period planning MINLP model for shale gas development projects is presented in the work by Drouven and Grossmann [17], which extends the work by Cafaro and Grossmann [13]. The model uses a shale gas development superstructure and optimizes decisions in the following areas:

- Selection of well pad locations,
- Timing of well pad construction,
- Selection of number of wells to drill and their location,
- Timing of drilling operations,
- Allocation of available drilling rigs,
- Selection of locations for processing plants and compressor stations,
- Allocation of available compressors, and
- Layout and design of pipelines.

A greenfield shale gas development case study was performed for a 10-year planning horizon with 10 well pad candidates in three pad clusters, 1 processing plant, 1 compressor station, 1 freshwater source, and varying well compositions by well clusters. The superstructure for this problem is the same as the one used for the case studies presented in the work by Drouven and Grossmann [17]. Figure 6 illustrates this superstructure. The resulting MINLP model consists of 13,000 constraints, 10,000 bilinear terms, and 25,000 binary variables. The model was optimized in approximately 1.5 h to yield an NPV of $214 million USD, which is 2.6 times higher than the historic NPV for that project ($81 million USD). This NPV was achieved by improving equipment utilization and scheduling return-to-pad operations, which incurred 14% more development expenses, but required 23% less wells to be drilled. The optimal structure and operating schedules are given in Fig. 7 and Fig. 8, respectively.

In the work by Cafaro, et al. [12], planning of shale gas refracturing is proposed via two main approaches, (1) a continuous-time NLP model based on productivity decline forecasts, and (2) a discrete-time MILP model that explicitly accounts for multiple refractures. The MILP model is obtained via reformulation of a GDP model, and takes three different forms depending on the reformulation approach (big-M reformulation, standard hull reformulation, and compact hull reformulation). The model is coupled with multiple price forecasting models and a reservoir simulation model with real data. The model sizes are given in Table 1. Although the hull reformulations result in significantly larger models, the model complexity is paid off with the superior computational performance observed. The reduced model size of the compact hull reformulation allows it to outperform the standard hull reformulation, taking less than half of the time required to solve the latter. For the

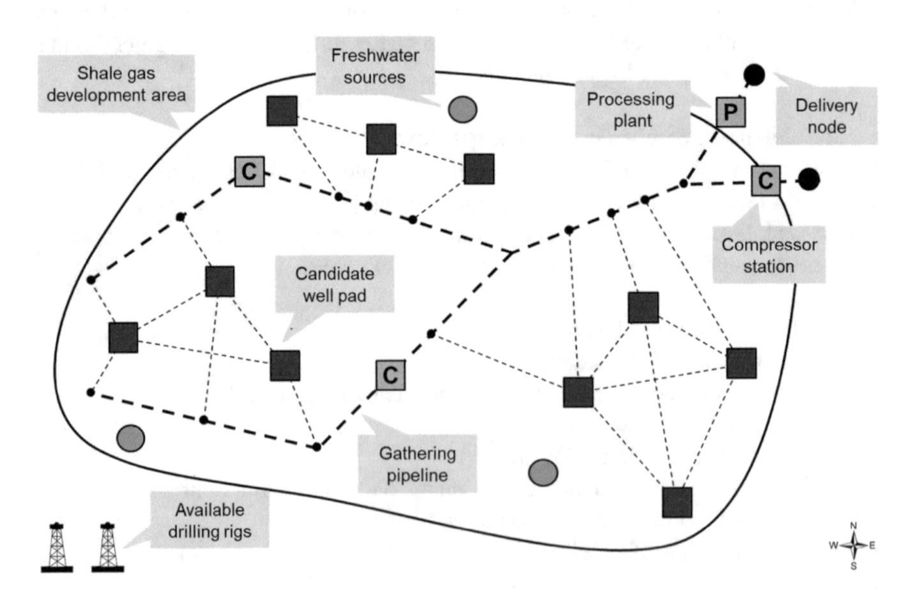

Fig. 6 Case study shale gas development superstructure (Adapted with permission from [17]. Copyright (2016) John Wiley & Sons Inc.)

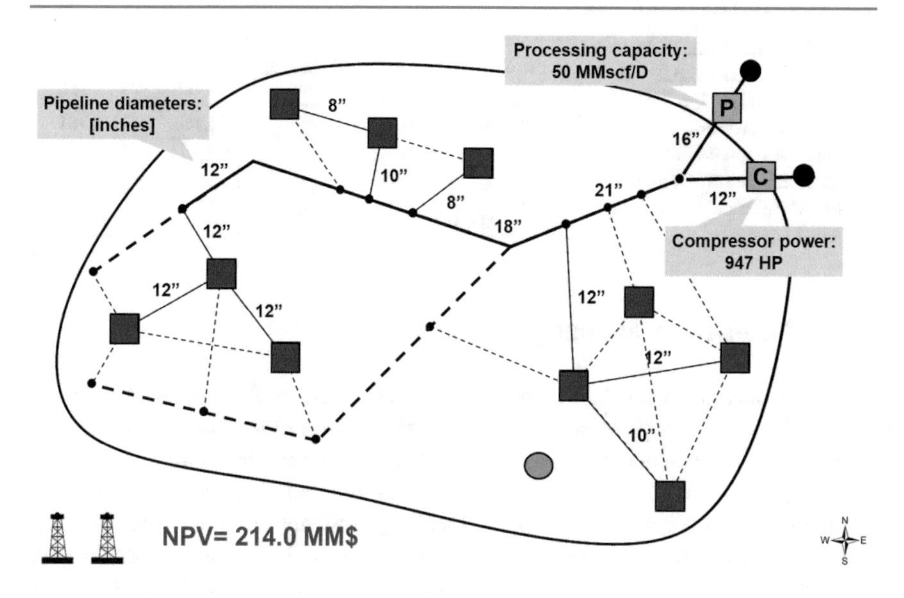

Fig. 7 Optimal shale gas structure for greenfield case study superstructure (Adapted with permission from [17]. Copyright (2016) John Wiley & Sons Inc.)

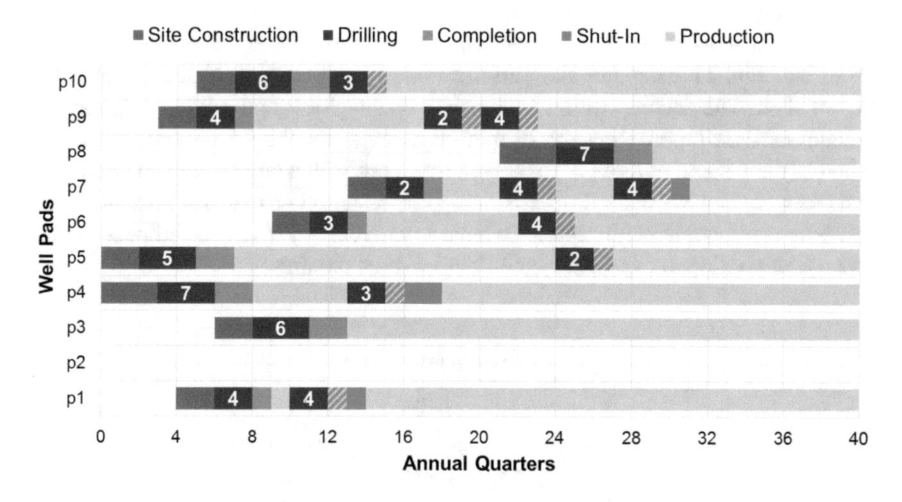

Fig. 8 Optimal shale gas development schedule for greenfield case study with number of wells drilled indicated within the drilling tasks

real case study, the solutions find significant increases in the development NPV and well recovery. Well recoveries are increased by up to 25% and profits by hundreds of thousands of dollars when using the optimized plan.

Table 1 Model sizes and run time (using Gurobi 5.6.2) for the big-M formulation (BMF), standard hull reformulation (SHR), and compact hull reformulation (CHR) MILP models [12]

	BMF	SHR	CHR
Binary variables	240	360	360
Continuous variables	481	44,161	22,381
Constraints	15,603	89,163	67,383
Nodes explored	7,786	0	0
Solution time (s)	255	22	9

4.2 Multi-period Blending

Blending operations are key in many industries, such as the downstream petrochemicals, food, and pharmaceuticals industries, where most product recipes contain one or more blending steps. Figure 9 shows a sample configuration for a blending problem consisting of supply, blending, and demand tanks. From a mathematical programming standpoint, blending problems are difficult to solve due to bilinear terms that give rise to non-convex MINLPs. Different approaches have been taken to improve the solvability of these models. In the work by Lotero, et al. [37], an alternate formulation with redundant constraints is used to tighten and improve the model relaxations. The authors also present a bilevel decomposition algorithm that outperforms state of the art general purpose solvers. In the decomposition algorithm, a master MILP is solved to fix binary variables in a subproblem containing a reduced MINLP. Figure 10 shows the proposed decomposition algorithm and Fig. 11 compares the performance of the proposed algorithm with the SCIP solver. The decomposition approach finds the global optimum in a reasonable amount of time (approximately 20 min on average for the 45 instances studied), whereas the general purpose SCIP solver, times out with a large optimality gap. The instances studied were for 6 and 8 time periods and 1–10 stream specifications, resulting in models with 240-320 binary variables, 128-1,760 bilinear terms, 552-1,312 continuous variables, and 984-3,616 constraints.

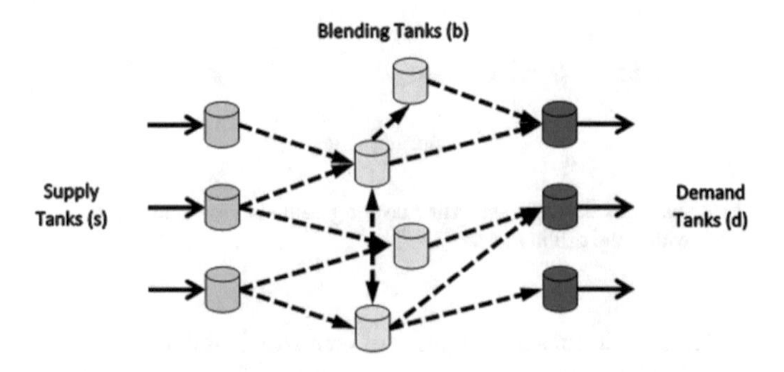

Fig. 9 Sample schematic for blending problem (Adapted with permission from [37]. Copyright (2016) Elsevier Ltd.)

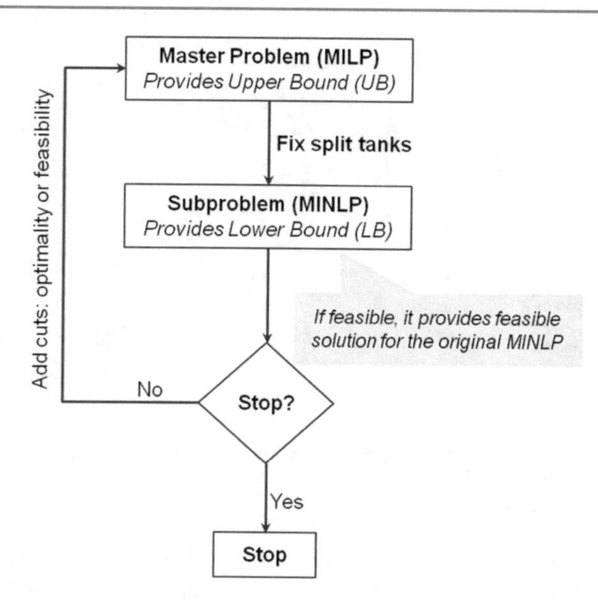

Fig. 10 Bilevel decomposition algorithm for multi-period blending problem (Adapted with permission from [37]. Copyright (2016) Elsevier Ltd.)

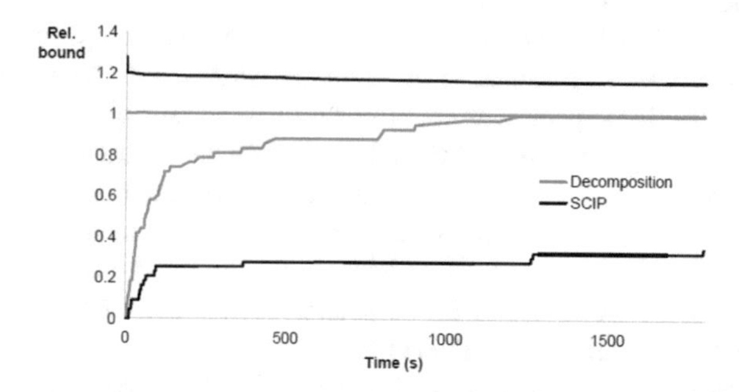

Fig. 11 43-instance average relative upper and lower bounds for decomposition algorithm and SCIP solver problem (Adapted with permission from [37]. Copyright (2016) Elsevier Ltd.)

4.3 Processing Plant Designs

MINLP has been applied to the optimization of biological and chemical processing plants. One example is the optimization of a natural gas plant design presented in the work by Caballero, et al. [11]. This work integrates GDP with the commercial process simulator HYSYS to find the optimal equipment types and utility selections for the plant. The reformulated MINLP is coupled to a HYSYS flowsheet and has 38 explicit nonlinear constraints, 5 linear constraints, 16 binary variables, 19 external variables (7 are flowsheet specifications), and 40 implicit blocks of

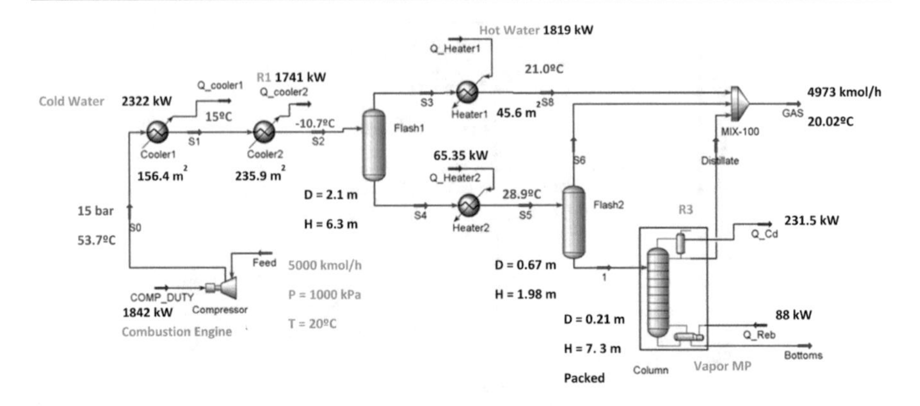

Fig. 12 Optimal natural gas plant design (Adapted with permission from [11]. Copyright (2007) John Wiley & Sons Inc.)

equations (for cost, sizing, and correlations). The two algorithms applied are an LP-NLP Branch-and-bound and Outer approximation. The LP-NLP BB algorithm solves 23 LP nodes and 2 NLP subproblems, requiring 300 s to find the minimum cost design of $117 thousand/year. The OA approach performs 4 major iterations and solves 3 NLP subproblems in 708 s. The optimal design with equipment selections and utility specifications is shown in Fig. 12.

Another example is the economic evaluation and design of a plant for bioethanol and biodiesel synthesis presented in the work by Martín and Grossmann [40]. The work uses MINLP for superstructure optimization of a plant with several alternative biochemical pathways. The model includes two biomass pretreatment alternatives, mainly ammonia fiber explosion (AFEX) and dilute acid pretreatment. Pretreatment is followed by hydrolysis, from which the intermediate material can either be fermented and purified to produce biodiesel or bioethanol. An additional feature of the model is that it accounts for heat integration for the plant by performing a heat exchanger network synthesis simultaneously during the optimization. From the data used in the model, the production of biodiesel is determined to be economically infeasible. The model allows to perform a sensitivity study to understand the technological improvements required in terms of conversion ratios (greater than 50%) for biodiesel to become economically feasible.

In another study by Martín and Grossmann [39], superstructure optimization with MINLP is used to design a lignocellulosic ethanol process via gasification. The superstructure includes alternate technological pathways for the different stages of bioethanol production as depicted in Fig. 13. The MINLP superstructure can be decomposed into 8 subproblems that can be solved as NLPs as shown in Fig. 14. Within each subproblem, the following subsystems are linked sequentially: syngas composition adjustment, sour gas removal, and ethanol purification. MINOS, KNITRO, and CONOPT3 (NLP solvers) are used to initialize the subsystems. Heat integration using the SYNHEAT heat integration software is then included on each of the subproblems to design an energetically optimal system. Multi-effect

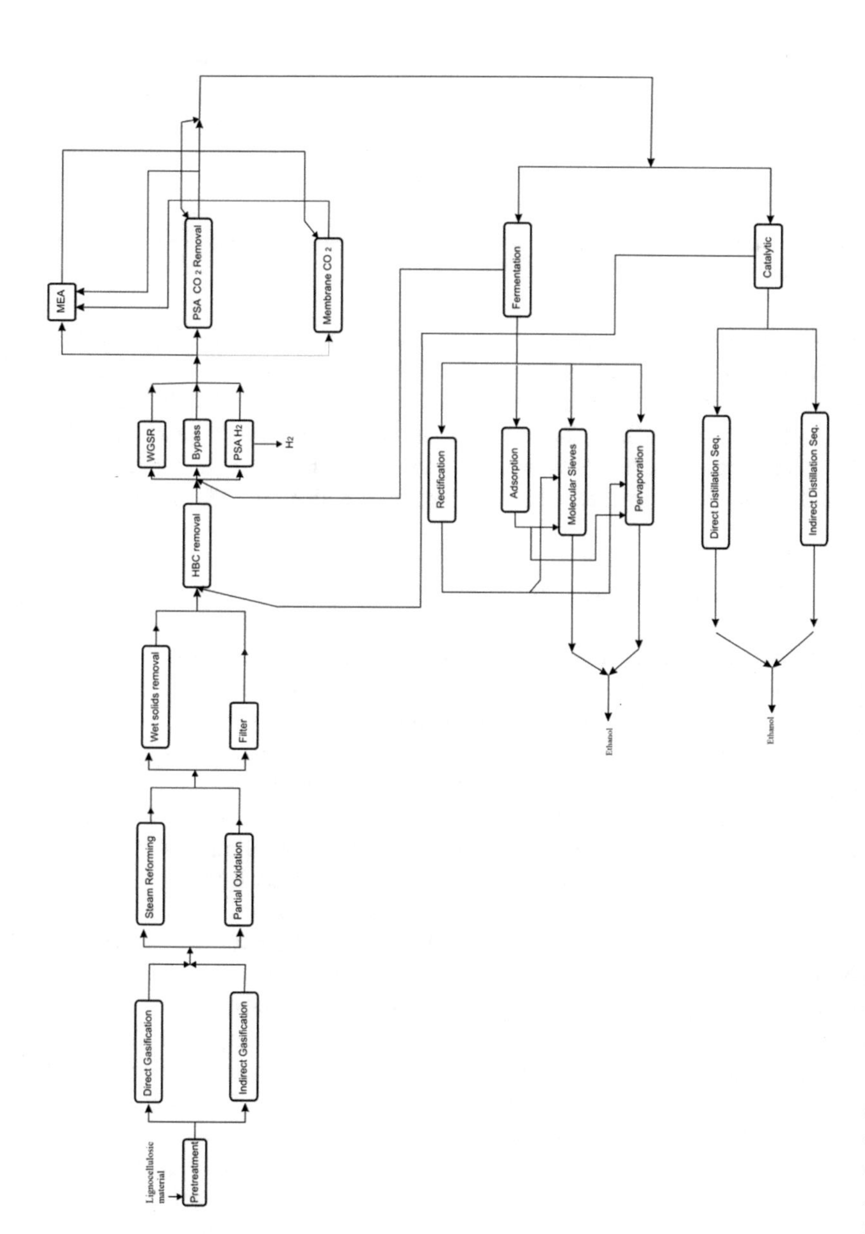

Fig. 13 Technological pathways for bioethanol production (Reproduced with permission from [39]. Copyright (2011) John Wiley & Sons Inc.)

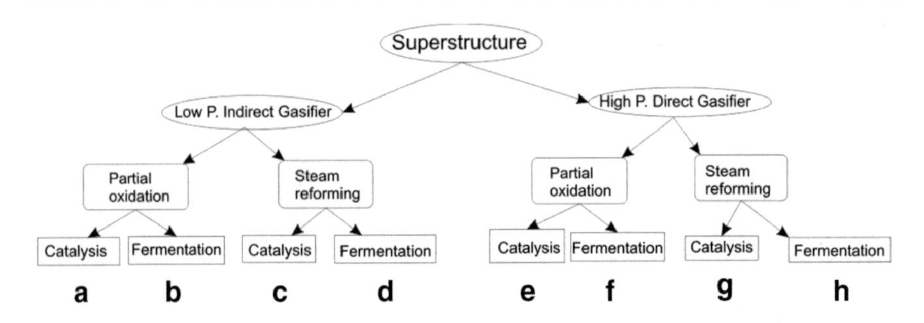

Fig. 14 Superstructure subproblems for bioethanol production (Reproduced with permission from [39]. Copyright (2011) John Wiley & Sons Inc.)

distillation is used to strengthen the heat integration in the final purification step. The optimal design results in a production cost of $1.04/gal of ethanol and involves high pressure gasification, steam reforming, PSA H_2, PSA/MEA, catalytic reactor, and direct distillation. The cost is further reduced with H_2 byproduct credits to $0.41/gal. These results show potential achieved through optimization when compared to the other cost evaluations in literature, which range from $1-2/gal.

4.4 Facility Network Design

The work by Lara, et al. [35] presents a novel algorithm for planning and designing manufacturing networks that takes into account the tradeoffs between the decision to build centralized facilities versus the decision to build distributed facilities. The tradeoff is of economies of scale versus transportation costs, which play key roles in the profitability of manufacturing networks. The model used is a multi-period GDP extension of the Capacitated Multi-facility Weber Problem that is reformulated as a non-convex MINLP and solved with an accelerated bilevel decomposition algorithm that provides tight bounds and outperforms the commercial global optimizers BARON, SCIP, and ANTIGONE as shown in Fig. 15. The model uses the locations of customers and suppliers and iteratively refines the grid partitioning of the two-dimensional geographical space to find the optimal facility locations and make the optimal facility type decisions to meet location specific customer demand and minimize costs.

An application is presented in the design of a bioethanol supply chain with 10 suppliers, 10 markets, 10 potential distributed facilities, and 2 potential centralized facilities. The model has 1,320 binary variables, 1,545 continuous variables, and 3,457 constraints. Three iterations of the accelerated bilevel decomposition algorithm during 6 h were required to solve the problem to a 2% optimality gap with a net present cost of approximately $2 billion. The proposed algorithm proved superior to BARON, which still had an optimality gap of 68% after 10 h of run time. The optimal supply chain design is illustrated in Fig. 16. The investment schedule includes building 10 distributed facilities in the first year and a single centralized facility in the second year.

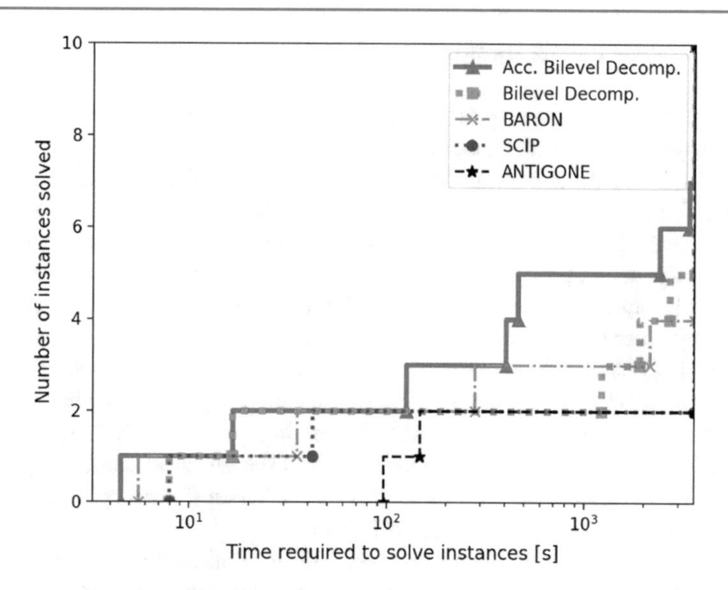

Fig. 15 Performance profiles for the standard and accelerated bilevel decomposition algorithm in the work by Lara, et al. [35, 36] and the commercial solvers BARON, SCIP, and ANTIGONE (Reproduced with permission from [35]. Copyright (2019) Elsevier Ltd.)

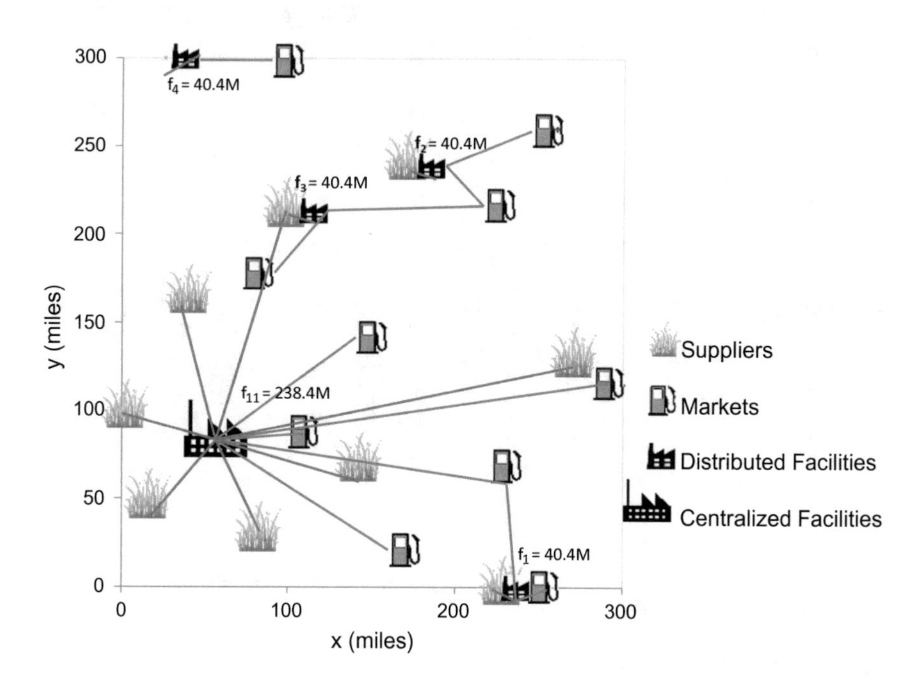

Fig. 16 Optimal supply chain network structure for the biomass case study (Reproduced with permission from [35]. Copyright (2019) Elsevier Ltd.)

4.5 Water Network Design

Global optimization techniques are important in solving water network design problems that require water treatment facilities for water leaving processing facilities, a common requirement in many industries. This problem is formulated in the work by Karuppiah and Grossmann [33] as a superstructure optimization with a non-convex NLP or a non-convex GDP, in which all possible interconnections between water-using process equipment and treatment units are considered so as to account for reuse and recycle of water. Techniques such as convex envelopes, explicit tight variable bounds inferred from the superstructure, and novel bound strengthening cuts based on flow balances are used in a spatial branch and contract algorithm to solve this problem efficiently. The strengthening cuts also prove very advantageous when used with BARON, solving problems that were virtually unsolvable by BARON. As an example, in one of the instances studied, no solution was found after more than 10 h of solution time. When the strengthening cut was added to BARON, the problem solved in 1.06 s.

The work by Karuppiah and Grossmann [33] is continued in that of Ahmetović and Grossmann [1]. An industrial water network case study is depicted with the superstructure in Fig. 17. The notation in this superstructure is as follows: nodes with *PU* are processing units (5 in total), nodes with *TU* are treatment units (3 in

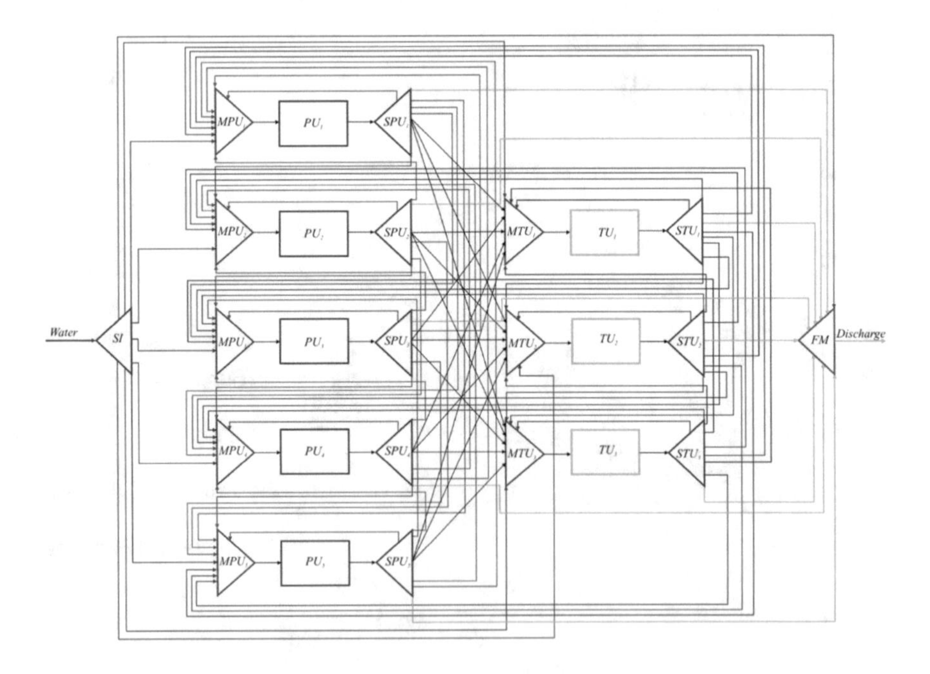

Fig. 17 Superstructure for facility water network in case study

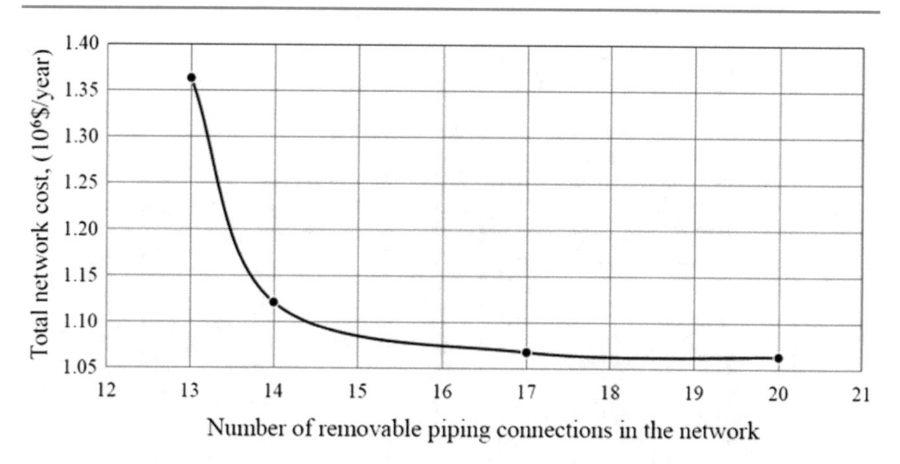

Fig. 18 Pareto-optimal front for the water network design case study (Reproduced with permission from [1]. Copyright (2011) John Wiley & Sons Inc.)

total), red triangle nodes are mixers, and blue triangle nodes are splitters. The study quantifies the tradeoff between investment cost and the network complexity. A pareto plot of this tradeoff is shown in Fig. 18. For all the points in this front, the water consumption was 40 t/h, representing a reduction of more than 85% in the standard water consumption at the plant. The simplest network, with 13 removable connections is depicted in Fig. 19. The model for this instance has 72 binary variables, 233 continuous variables, and 251 constraints. Solution time in BARON is under 200 s with a 1% optimality gap stopping criteria.

Other works have also shown that these design problems can be solved with increased levels of detail in the process models used. The work by Yang, et al. [49] introduces shortcut models for the treatment units rather than using simple fixed contaminant removal models. This approach makes optimization results more realistic and applicable. The shortcut models used, account for mass transfer of contaminants in the treatment units. This allows the optimization models to represent the connection between treatment costs and removal efficiencies, providing more realistic solutions. Other extensions include the use of nonlinear cost functions for operating and investment costs of treatment units, and addressing uncertainty in the contaminant loads present via worst-case, best-case, and nominal scenarios. The increased complexity of the models is mitigated by using Lagrangean decomposition strategies to solve the problems to global optimality.

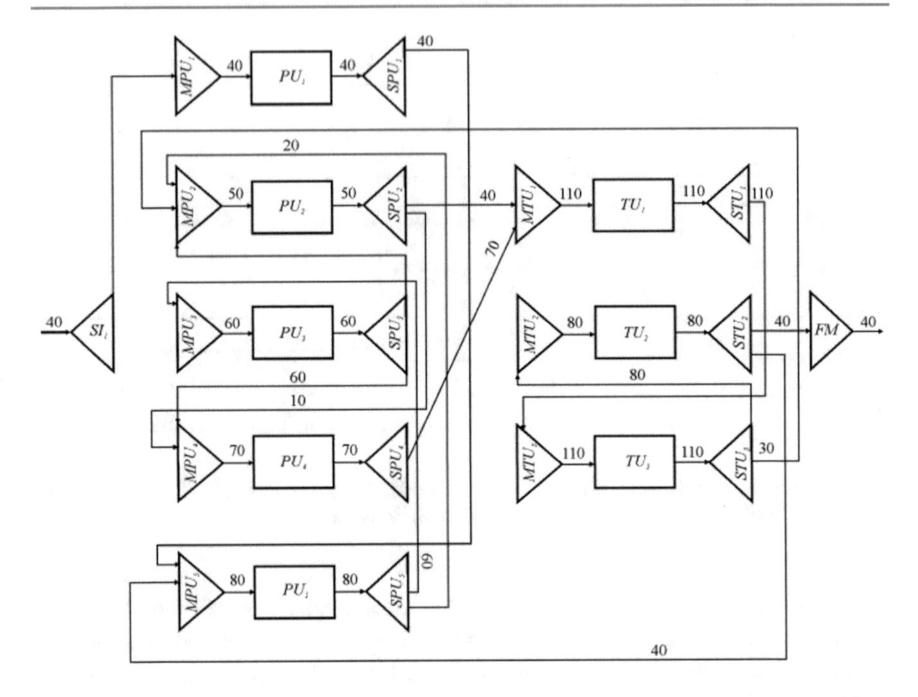

Fig. 19 Simplest configuration on the pareto front for the optimal water network design (Reproduced with permission from [1]. Copyright (2011) John Wiley & Sons Inc.)

4.6 Electricity Market Integration

Cryogenic energy storage (CES) technology makes it possible for chemical industries to integrate their processes with the energy market. In the work by Zhang, et al. [51], an optimization model is proposed to integrate the operations of an air separation unit (ASU) with the energy market via CES. The process, which is illustrated in Fig. 20, allows for an ASU to produce surplus liquid nitrogen and oxygen to create a CES inventory that can then be used for one or more of the following alternatives: (1) to generate electricity with a gas turbine for internal use, (2) to sell electricity at the spot price on the energy market, or (3) to commit electricity on the reserve market. High electricity spot prices make options 1 and 2 attractive, whereas the fact that revenue from the operating reserve market is created regardless of whether the electricity is actually sold or how much of it is sold makes option 3 attractive. However, option 3 introduces uncertainty that must be properly addressed to avoid severe penalties for not providing the reserve capacity immediately upon request. The uncertainty present here is three-fold: amount, time, and

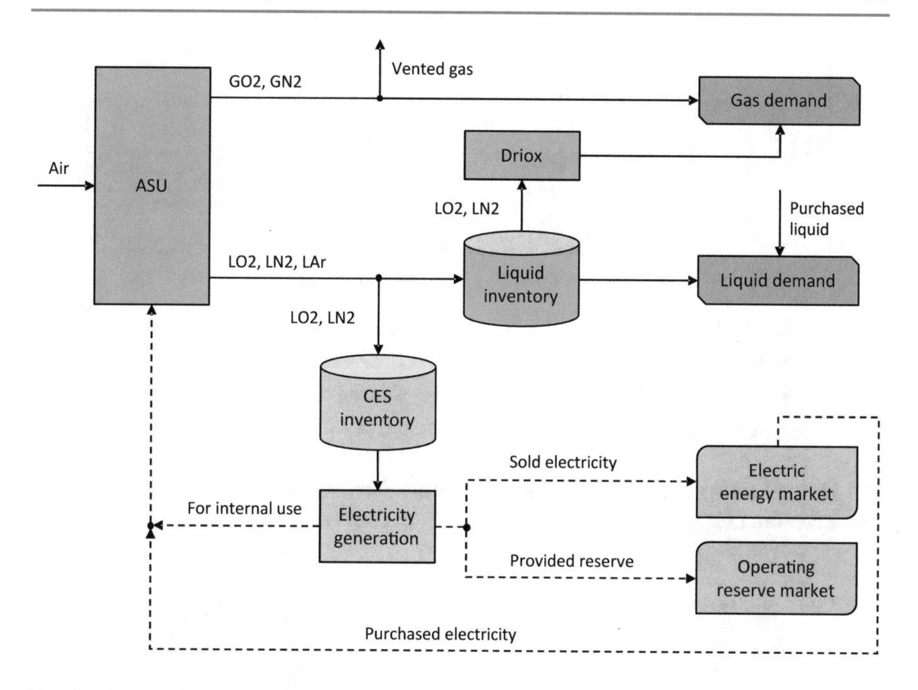

Fig. 20 Proposed ASU/CSE system for energy market integration (Reproduced with permission from [51]. Copyright (2015) John Wiley & Sons Inc.)

duration of the reserve demand. To properly plan for such uncertainty and capture the benefits of participating in the operating reserve market, adjustable affine robust optimization (AARO) is used on an MILP scheduling model for the integrated ASU/CES system. The model has 2.5 thousand binary variables, 55 thousand continuous variables, and 53 thousand constraints. Solution time with CPLEX 12.5 is 10 min with a 1% optimality gap stopping criteria.

The proposed ASU/CES system is tested under varying degrees of conservatism in regard to the operating reserve demand. Even with the highest level of conservatism (assuming that operating reserve can be requested every day), a 5% cost reduction in utilities, especially during peak times, is achieved under the proposed system. For moderate levels of conservatism, the savings are on the order of 9%. Figure 21 shows the optimal operating schedule for the most conservative case. Another benefit of the integrated ASU/CES system is that of increased plant operation as shown in Fig. 22.

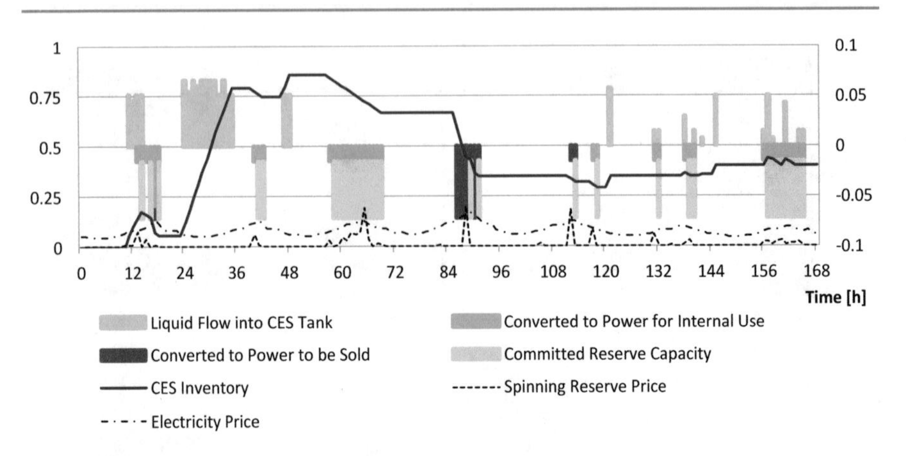

Fig. 21 Optimal operating schedule for conservative reserve demand scenario (Reproduced with permission from [51]. Copyright (2015) John Wiley & Sons Inc.)

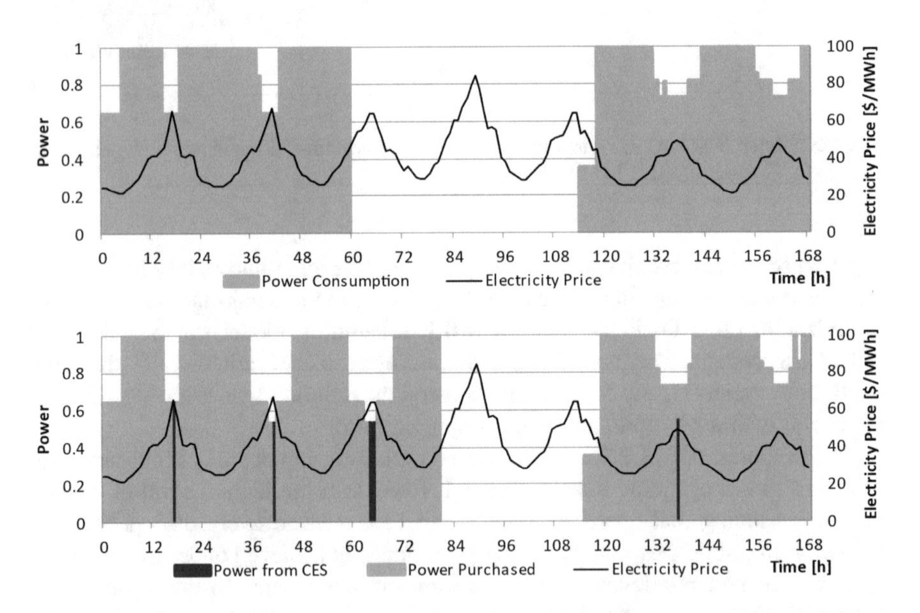

Fig. 22 Normalized power and electricity price profiles for the design with no CES capabilities (top) and the design with CES (bottom) (Reproduced with permission from [51]. Copyright (2015) John Wiley & Sons Inc.)

4.7 Reliable Plant Design

In the chemical and manufacturing industries, plant reliability is essential for maintaining competitivity and profitability. However, plant synthesis and optimization approaches often ignore the need for redundant equipment and

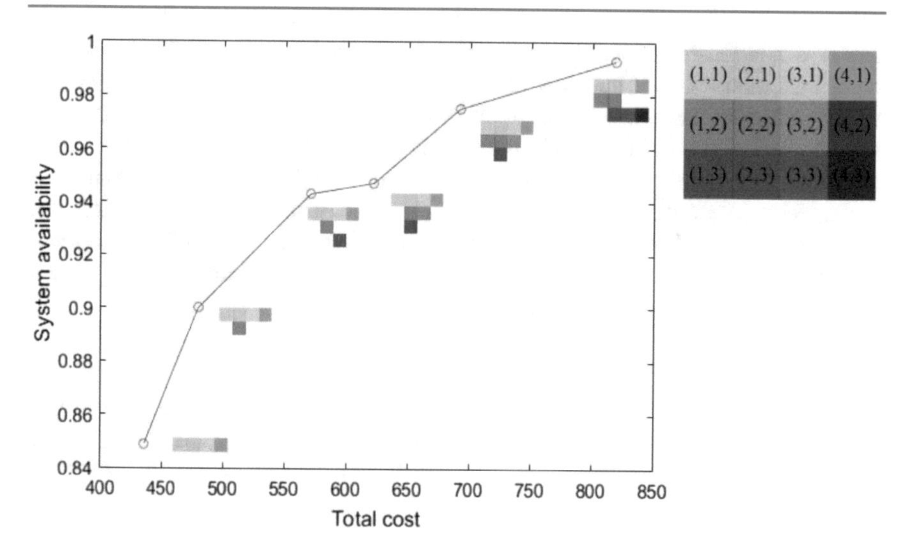

Fig. 23 Sample pareto plot of system availability versus total equipment cost for a system with four stages and three potential redundant units per stage (Reproduced with permission from [50]. Copyright (2018) Elsevier Ltd.)

probabilities of equipment failure. The work by Ye, et al. [50] proposes a model for designing reliable plants that are serial in structure. Options within the superstructure include installation of prioritized redundant equipment with identical characteristics and redundant equipment with varying capacities or features. A geometric distribution is used to model stage availability. Equipment costs in the model include both installation and repair costs. Two formulations are presented with different objectives: (1) profit maximization and (2) plant availability maximization. An advantage of the second formulation is that it can be reformulated as a convex MINLP, whereas the profit maximization formulation is non-convex. The availability maximization model uses an ϵ-constraint to limit equipment costs. The model facilitates assessing the tradeoffs between plant availability and cost as shown in Fig. 23. For designs below the pareto curve, it is possible to simultaneously improve availability and reduce cost up until a point on the pareto curve is attained. However, beyond that point, it is not possible to simultaneously improve both objectives. An improvement in one objective, requires a worsening in the other. The only way to improve availability is to increase the upper bound on the equipment cost.

Two case studies for the reliable design of a methanol synthesis plant [47] and a toluene hydrodealkylation plant [34] are conducted. The methanol synthesis reliability model consists of 72 binary variables, 451 continuous variables, and 408 constraints. Solution time is 0.45 s with DICOPT (CONOPT 3.16D and CPLEX 12.6) to attain an optimal profit of \$3.40 million/yr and an availability of 97%. This is an improvement in the optimal design presented in the literature, which ignores potential equipment failures, resulting in a profit of \$3.35 million/yr (1.5% lower) and an availability of 92%. The hydrodealkylation model is larger with 400 million design alternatives, 108 binary variables, 955 continuous variables, and 893 constraints, yielding an optimal profit of $3.97 million/yr in 10 s of solution time. This solution has a profit that is 4% higher than a naïve design that ignores plant reliability. Availability is increased from 90 to 94% by considering equipment redundancy. The optimal reliable designs for the methanol and hydrodealkylation plants are given in Figs. 24 and 25. Reliability optimization allows for increases in both plant reliability and profit in the order of a few percentage points since it mitigates losses due to plant failures.

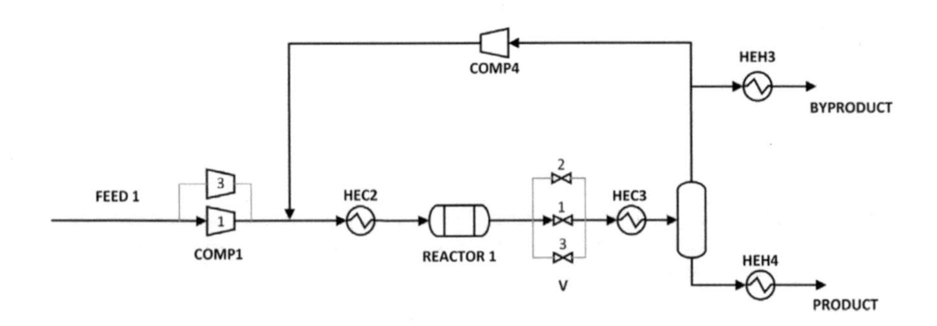

Fig. 24 Optimal design for a methanol synthesis plant [47] with redundant units (Reproduced with permission from [50]. Copyright (2018) Elsevier Ltd.)

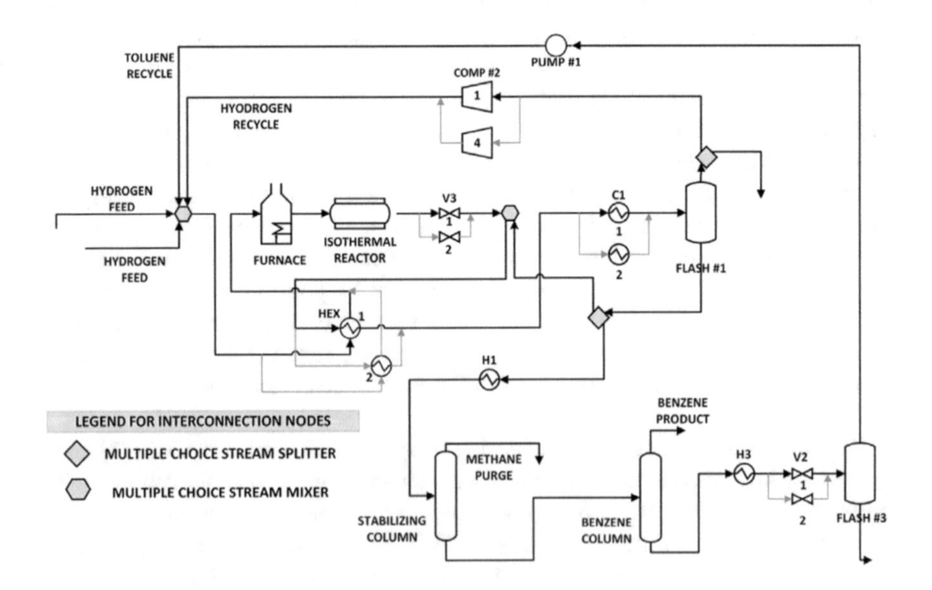

Fig. 25 Optimal design for a hydrodealkylation plant [34] with redundant units (Reproduced with permission from [50]. Copyright (2018) Elsevier Ltd.)

4.8 Resilient Supply Chain Design

Expanding the scope of reliability to the supply chain level, reliability of supply chains in view of disturbances is critical to effective supply chain management. The motivating problem in the work by Garcia-Herreros, et al. [20] is that of a supply chain with one manufacturing facility that produces multiple commodities that are sent to multiple distribution centers to satisfy demands at different customer locations (Fig. 26). Potential distribution center locations are preselected. Disturbances in the system occur when disruptions occur at the distribution centers. There is a fixed probability of disruption for each distribution. The key decisions are whether to establish a distribution center at a preselected location and with what capacity to install it. A two-stage stochastic programming model is proposed, where the first stage decisions are the distribution center selection and capacity assignments, and the second stage decisions are the assignment of commodity specific customer demands to each available distribution center under each disruption scenario. The model is solved with a strengthened multi-cut Benders decomposition [10]. Additionally, several techniques are applied to improve the model's solvability. The four main techniques are as follows:

1. Indistinguishability: indistinguishable scenarios are identified as those with disruptions at locations that were not selected. Only one instance is solved for each indistinguishable set.
2. Parallelization: Benders subproblems are parallelized.
3. Relevant scenario selection: scenarios with very low probabilities of occurring are set aside to allow the model to be tractable despite the exponential increase in scenarios as the problem size increases.
4. Bounding excluded scenarios: a procedure for providing bounds on the full-space model that accounts for the excluded scenarios after the optimization completes is used.

The model is applied to an industrial supply chain design problem with 29 candidate distribution centers with disruption probabilities ranging from 0.5 to 3%, 110 customers, and 61 commodities. The total number of scenarios that would need to be considered is $2^{29} \approx 540\,million$. However, using the techniques described previously, the problem can be solved in under 5 h for a reduced problem set with at most 1 disruption at a time and 30 scenarios that covers 85% of the total probability of scenarios. The model size for this instance is of 29 binary variables, 250 thousand continuous variables, and 300 thousand constraints. Increasing the coverage to 98.5% of the total probability of scenarios (436 scenarios in the set), by allowing up to 2 simultaneous disruptions, increases the run time to around 5 days with less than a 1% optimality gap. The model size here is 29 binary variables, 3.6 million continuous variables, and 4.4. million constraints. In this case, the scenario set is within the full-space bounds, which is not the case in the deterministic problem nor in the 30-scenario set problem.

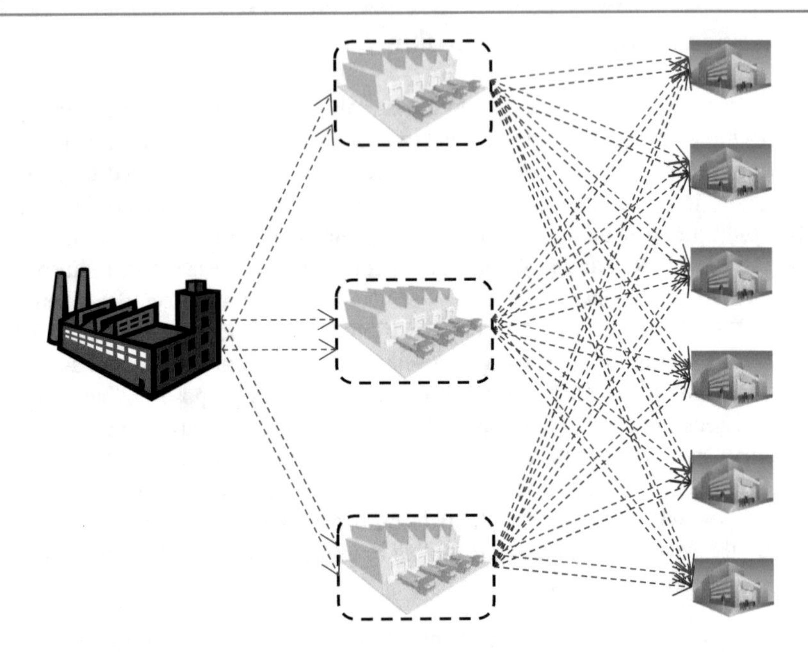

Fig. 26 Sample multi-commodity distribution system with 1 plant, 3 distribution centers, and 6 customers (Adapted with permission from [20]. Copyright (2014) American Chemical Society)

5 Conclusions

An overview of continuous and discrete optimization models has been presented with successful applications of these models in industrial settings. The industrial applications include Oil & Gas upstream operations, material blending facilities, natural gas plant design, biofuels synthesis, supply chain facility network design, water network design, industrial electricity market integration, reliable plant design, and supply chain design. Strategies for improving model tractability have been presented, including model linearization, decomposition methods, scenario subset selection, and strengthening cuts, among others. Continuous and discrete optimization in industrial applications remains a rich and thriving field of research with new problems being engaged every year. Algorithmic improvements over the years have made it possible to solve larger models. Some key challenges that are being and need to be tackled are: (1) finding improved relaxations for global optimization, (2) improving the computational performance of large-scale non-convex MINLP/GDP and stochastic programming, (3) developing algorithms for mixed-integer dynamic optimization that enable integration among the three levels of the decision-making pyramid, (4) optimizing entire supply chains, and (5) extending the work to include sustainable system design and operation. Further developments in these areas will greatly improve the quality of decisions that need to be made in the processing industries.

Acknowledgements The authors acknowledge the support of the Center for Advanced Process Decision-making (CAPD) at Carnegie Mellon University and each of the CAPD and EWO member companies.

References

1. Ahmetović, E., Grossmann, I.E.: Global superstructure optimization for the design of integrated process water networks. AIChE J. **57**, 434–457 (2011). https://doi.org/10.1002/aic.12276
2. Balas, E., Ceria, S., Cornuéjols, G.: A lift-and-project cutting plane algorithm for mixed 0-1 programs. Math. Program. **58**, 295–324 (1993). https://doi.org/10.1007/BF01581273
3. Beale, E.M.L.: Survey of integer programming. OR **16**, 219 (1965). https://doi.org/10.2307/3007503
4. Benders, J.F.: Partitioning procedures for solving mixed-variables programming problems. Numer. Math. **4**, 238–252 (1962). https://doi.org/10.1007/BF01386316
5. Ben-Tal, A., Goryashko, A., Guslitzer, E., Nemirovski, A.: Adjustable robust solutions of uncertain linear programs. Math. Program. **99**, 351–376 (2004). https://doi.org/10.1007/s10107-003-0454-y
6. Bertsimas, D., Brown, D.B., Caramanis, C.: Theory and applications of robust optimization. SIAM **53**, 464–501 (2011). https://doi.org/10.1137/080734510
7. Bertsimas, D., Sim, M.: Robust discrete optimization and network flows. In: Mathematical Programming, pp 49–71. Springer (2003)
8. Biegler, L.T.: Nonlinear Programming: Concepts, Algorithms, and Applications to Chemical Processes. SIAM (2010)
9. Birge, J., Louveaux, F.: Introduction to Stochastic Programming, 2nd edn. Springer Science & Business Media (2011)
10. Birge, J.R., Louveaux, F.V.: A multicut algorithm for two-stage stochastic linear programs. Eur. J. Oper. Res. **34**, 384–392 (1988). 10.1016/0377-2217(88)90159-2
11. Caballero, J.A., Odjo, A., Grossmann, I.E.: Flowsheet optimization with complex cost and size functions using process simulators. AIChE J. **53**, 2351–2366 (2007). https://doi.org/10.1002/aic.11262
12. Cafaro, D.C., Drouven, M.G., Grossmann, I.E.: Optimization models for planning shale gas well refracture treatments. AIChE J. **62**, 4297–4307 (2016). https://doi.org/10.1002/aic.15330
13. Cafaro, D.C., Grossmann, I.E.: Strategic planning, design, and development of the shale gas supply chain network. AIChE J. **60**, 2122–2142 (2014). https://doi.org/10.1002/aic.14405
14. Charnes, A., Cooper, W.W., Mellon, B.: Blending aviation gasolines–a study in programming interdependent activities in an integrated oil company. Econometrica **20**, 135 (1952). https://doi.org/10.2307/1907844
15. Cooper, W.W., Charnes, A., Cooper, W.W. et al.: A brief history of a long collaboration in developing industrial uses of linear programming. Oper. Res. **50**, 35–41 (2002)
16. Dakin, R.J.: A tree-search algorithm for mixed integer programming problems. Comput. J. **8**, 250–255 (1965). https://doi.org/10.1093/comjnl/8.3.250
17. Drouven, M.G., Grossmann, I.E.: Multi-period planning, design, and strategic models for long-term, quality-sensitive shale gas development. AIChE J. **62**, 2296–2323 (2016). https://doi.org/10.1002/aic.15174
18. Duran, M.A., Grossmann, I.E.: An outer-approximation algorithm for a class of mixed-integer nonlinear programs. Math. Program. **36**, 307–339 (1986). https://doi.org/10.1007/BF02592064
19. Escudero, L.F., Garín, A., Merino, M., Pérez, G.: The value of the stochastic solution in multistage problems. TOP **15**, 48–64 (2007). https://doi.org/10.1007/s11750-007-0005-4

20. Garcia-Herreros, P., Wassick, J.M., Grossmann, I.E.: Design of resilient supply chains with risk of facility disruptions. Ind. Eng. Chem. Res. **53**, 17240–17251 (2014). https://doi.org/10.1021/ie5004174
21. Geoffrion, A.M.: Generalized benders decomposition. J. Optim. Theory Appl. **10**, 237–260 (1972). https://doi.org/10.1007/BF00934810
22. Grossmann, I.: Enterprise-wide optimization: a new frontier in process systems engineering. AIChE J., 1846–1857 (2005)
23. Grossmann, I.E., Calfa, B.A., Garcia-Herreros, P.: Evolution of concepts and models for quantifying resiliency and flexibility of chemical processes. Comput. Chem. Eng. **70**, 22–34 (2014). https://doi.org/10.1016/j.compchemeng.2013.12.013
24. Grossmann, I.E., Halemane, K.P., Swaney, R.E.: Optimization strategies for flexible chemical processes. Comput. Chem. Eng. **7**, 439–462 (1983). https://doi.org/10.1016/0098-1354(83)80022-2
25. Grossmann, I.E., Trespalacios, F.: Systematic modeling of discrete-continuous optimization models through generalized disjunctive programming. AIChE J. **59**, 3276–3295 (2013). https://doi.org/10.1002/aic.14088
26. Gupta, V., Grossmann, I.E.: An efficient multiperiod MINLP model for optimal planning of offshore oil and gas field infrastructure. Ind. Eng. Chem. Res. **51**, 6823–6840 (2012). https://doi.org/10.1021/ie202959w
27. Gupta, V., Grossmann, I.E.: Multistage stochastic programming approach for offshore oilfield infrastructure planning under production sharing agreements and endogenous uncertainties. J. Petrol. Sci. Eng. **124**, 180–197 (2014). https://doi.org/10.1016/j.petrol.2014.10.006
28. Hooker, J.N., van Hoeve, W.J.: Constraint programming and operations research. Constraints **23**, 172–195 (2018). https://doi.org/10.1007/s10601-017-9280-3
29. Illés, T., Terlaky, T.: Pivot versus interior point methods: pros and cons. Eur. J. Oper. Res., 170–190 (2002). North-Holland
30. Jain, V., Grossmann, I.E.: Algorithms for hybrid MILP/CP Models For A Class Of Optimization Problems. INFORMS J. Comput. **13**, 258–276 (2001). https://doi.org/10.1287/ijoc.13.4.258.9733
31. Johnson, E.L., Nemhauser, G.L., Savelsbergh, M.W.P.: Progress in Linear programming-based algorithms for integer programming: an exposition. INFORMS J. Comput. **12**, 2–23 (2000). https://doi.org/10.1287/ijoc.12.1.2.11900
32. Jonsbråten, T.W.: Optimization models for petroleum field exploitation. PhD thesis, NHH Norwegian School of Economics and Business Administration, Bergen, Norway (1998)
33. Karuppiah, R., Grossmann, I.E.: Global optimization for the synthesis of integrated water systems in chemical processes. Comput. Chem. Eng. **30**, 650–673 (2006). https://doi.org/10.1016/j.compchemeng.2005.11.005
34. Kocis, G.R., Grossmann, I.E.: A modelling and decomposition strategy for the MINLP optimization of process flowsheets. Comput. Chem. Eng. **13**, 797–819 (1989). https://doi.org/10.1016/0098-1354(89)85053-7
35. Lara, C.L., Bernal, D.E., Li, C., Grossmann, I.E.: Global optimization algorithm for multi-period design and planning of centralized and distributed manufacturing networks. Comput. Chem. Eng. **127**, 295–310 (2019). https://doi.org/10.1016/j.compchemeng.2019.05.022
36. Lara, C.L., Trespalacios, F., Grossmann, I.E.: Global optimization algorithm for capacitated multi-facility continuous location-allocation problems. J. Global Optim. **71**, 871–889 (2018). https://doi.org/10.1007/s10898-018-0621-6
37. Lotero, I., Trespalacios, F., Grossmann, I.E., Papageorgiou, D.J., Cheon, M.S.: An MILP-MINLP decomposition method for the global optimization of a source based model of the multiperiod blending problem. Comput. Chem. Eng. **87**, 13–35 (2016). https://doi.org/10.1016/j.compchemeng.2015.12.017

38. Maravelias, C.T., Grossmann, I.E.: A hybrid MILP/CP decomposition approach for the continuous time scheduling of multipurpose batch plants. Comput. Chem. Eng. **28**, 1921–1949 (2004). https://doi.org/10.1016/j.compchemeng.2004.03.016
39. Martín, M., Grossmann, I.E.: Energy optimization of bioethanol production via gasification of switchgrass. AIChE J. **57**, 3408–3428 (2011). https://doi.org/10.1002/aic.12544
40. Martín, M., Grossmann, I.E.: Optimal simultaneous production of biodiesel (FAEE) and bioethanol from switchgrass. Ind. Eng. Chem. Res. **54**, 4337–4346 (2015). https://doi.org/10.1021/ie5038648
41. Mieles, C.: Global Oil & Gas Exploration & Production (2020)
42. Powell, M.J.D.: Evelyn Martin Lansdowne Beale. 8 September 1928–23 December 1985. Biograp. Mem. Fellows R. Soc. **33**, 23–45 (1987)
43. Raman, R., Grossmann, I.E.: Modelling and computational techniques for logic based integer programming. Comput. Chem. Eng. **18**, 563–578 (1994). https://doi.org/10.1016/0098-1354(93)E0010-7
44. Sahinidis, N.V.: Optimization under uncertainty: state-of-the-art and opportunities. In: Computers and Chemical Engineering, pp. 971–983. Pergamon (2004)
45. Su, L., Tang, L., Bernal, D.E., Grossmann, I.E.: Improved quadratic cuts for convex mixed-integer nonlinear programs. Comput. Chem. Eng. **109**, 77–95 (2018). https://doi.org/10.1016/j.compchemeng.2017.10.011
46. Tomlin, J.A.: A note on comparing simplex and interior methods for linear programming. Progress in Mathematical Programming—Interior-Point and Related Methods, pp. 91–103. Springer, New York (1989)
47. Türkay, M., Grossmann, I.E.: Logic-based MINLP algorithms for the optimal synthesis of process networks. Comput. Chem. Eng. **20**, 959–978 (1996). https://doi.org/10.1016/0098-1354(95)00219-7
48. Westerlund, T., Pettersson, F.: An extended cutting plane method for solving convex MINLP problems. Comput. Chem. Eng. **19**, 131–136 (1995). https://doi.org/10.1016/0098-1354(95)87027-X
49. Yang, L., Salcedo-Diaz, R., Grossmann, I.E.: Water network optimization with wastewater regeneration models. Ind. Eng. Chem. Res. **53**, 17680–17695 (2014). https://doi.org/10.1021/ie500978h
50. Ye, Y., Grossmann, I.E., Pinto, J.M.: Mixed-integer nonlinear programming models for optimal design of reliable chemical plants. Comput. Chem. Eng. **116**, 3–16 (2018). https://doi.org/10.1016/j.compchemeng.2017.08.013
51. Zhang, Q., Grossmann, I.E., Heuberger, C.F., Sundaramoorthy, A., Pinto, J.M.: Air separation with cryogenic energy storage: optimal scheduling considering electric energy and reserve markets. AIChE J. **61**, 1547–1558 (2015). https://doi.org/10.1002/aic.14730
52. Zhang, Q., Grossmann, I.E., Lima, R.M.: On the relation between flexibility analysis and robust optimization for linear systems. AIChE J. **62**, 3109–3123 (2016). https://doi.org/10.1002/aic.15221

Optimal Design of a Railway Bypass at Parga, Northwest of Spain

Gerardo Casal, Alberte Castro, Duarte Santamarina and Miguel E. Vázquez-Méndez

Abstract

In the past few years, different models have been proposed to obtain the optimal design of a linear infrastructure (road or railway) connecting two given points. This paper analyses the usefulness of one of these models to design a railway bypass, in a real case study, on the railway line A Coruña-Palencia (Spain), where it passes through the urban area of Parga. Firstly, to show the good performance of the model, it is applied to a situation whose solution is known "a priori". Next, the problems arising in the real case are shown, where it is quite complicated to design a layout avoiding existing buildings and other restricted areas, and a two-stage method is proposed generating the layout of the requested bypass. Throughout the development of this method, it arises the need to connect a given circular curve with an also given tangent. To address this problem, an algorithm is proposed for computing the transition curve (clothoid) performing that connection. The work ends with some interesting conclusions and a brief description of future work.

G. Casal · D. Santamarina
Departamento de Matemática Aplicada, Universidade de Santiago de Compostela, Escola Politécnica Superior de Enxeñaría, 27002 Lugo, Spain
e-mail: xerardo.casal@usc.es

D. Santamarina
e-mail: duarte.santamarina@usc.es

A. Castro
Departamento de Enxeñaría Agroforestal, Universidade de Santiago de Compostela, Escola Politécnica Superior de Enxeñaría, 27002 Lugo, Spain
e-mail: alberte.castro@usc.es

M. E. Vázquez-Méndez (✉)
Departamento de Matemática Aplicada, Instituto de Matemáticas, Universidade de Santiago de Compostela, Escola Politécnica Superior de Enxeñaría, 27002 Lugo, Spain
e-mail: miguelernesto.vazquez@usc.es

P. Quintela Estévez et al. (eds.), *Advances on Links Between Mathematics and Industry*, SxI - Springer for Innovation / SxI - Springer per l'Innovazione 15,
https://doi.org/10.1007/978-3-030-59223-3_2

1 Introduction

Level crossings (crossings or intersections at the same level between a road and a railway) constitute a serious safety problem in all countries throughout the world. In Spain, the construction of new level crossings has been forbidden since 1978, and since 1987 the intersection between a road and a railway generated due to an establishment or modification of a new layout (either road or railway), it must be, mandatory built, at a different level [2]. In spite of that, in 30 June 2004, the number of level crossings in the Spanish railway network were still 4,465, out of which 3,737 were on public roads in service [2]. To reduce their number, a policy has been defined by the Spanish administration in order to achieve the progressive suppression of level crossings focusing on those with higher traffic density and train speed. To this aim, the Spanish Administrator of Railway Infrastructures (ADIF, from its Spanish acronym) is carrying out numerous actions in the past decades. As a general rule, these interventions consist of fencing the railway and modifying the infrastructure that goes through, by replacing it with an underpass or an overpass. These actions can deteriorate socially and economically the area, and on some occasions, they are firmly rejected by neighbours and local institutions. When this occurs, putting the railway track underground or constructing a bypass avoiding the urban area are usually the solutions most demanded by the population. These actions are much less aggressive at a social level, but they are more expensive in economic terms. In this scenario, as in many other life situations where social and economic aspects collide, the problem is to find a good solution balancing both aspects.

In recent years, numerous works have searched for the optimal layout linking two terminals, either by a road (see, for example, [3], [5] and its references therein), either by a railway (see, for example, [4] and [6]). In [9], the authors developed a model to optimize infrastructure costs in road design, and applied it to obtain a bypass on a Spanish national road. The optimal layout of a bypass in a railway is, however, a much less studied topic. Unlike what happens on a road, in a railway the bypass must connect correctly with the existing infrastructure geometry and, in addition, be designed with horizontal curves of greater radius to achieve the desired operational speed of the line, which makes the task of dodge the existing obstacles harder. This work focuses on the search for an optimal bypass, in economic terms, on the rail line A Coruña-Palencia (Spain) encircling the urban area of Parga and avoiding all buildings in the surrounding area. The work has been developed at the proposal of the movement #PARGANONSEDIVIDE, founded to disagree with the plan given by ADIF to suppress the local level crossings. In Sect. 2, the work under study is detailed, and following [9], it is formulated as a non-convex constrained optimization problem. Then, in Sect. 3, two different numerical experiments are developed: the first one shows the good behaviour of the model, while the second problem attempts to fulfil the needs defended by the neighbourhood movement. To obtain the requested bypass, a two-stage method is proposed. During the development of this method, the need arises to connect a given circular curve with a straight segment also given.

For this aim, an algorithm computing the transition curve enabling that connection is proposed (see Appendix). The paper ends (Sect. 4) with some interesting conclusions and a brief outline of future work.

2 Case Study and Mathematical Formulation

Parga is a village sited in the northwest of Spain. The village has grown along the railway line connecting A Coruña with Palencia (see Fig. 1). Nowadays, in the urban area there are three level crossings with different protection systems (two of them for vehicles and other for pedestrians), which ADIF intends to remove, by fencing the track area and building an underground passage in the centre of the village for all types of traffic. In response, the residents, who believe that this intervention would split the village into half, assemble to create the organization #PARGANONSEDIVIDE, to demand the construction of a railway bypass bordering the urban zone and dodge the buildings in the surrounding area. The problem that arises is the design of a bypass accomplishing this civil movement requirements, satisfying all technical standards, and being economically viable (optimally). Taking into account that any linear infrastructure is unequivocally determined by its horizontal alignment, its vertical alignment, and their cross-sections, this problem can be formulated as the following optimization problem (see [9]):

$$\min_{\mathbf{u}=(\mathbf{x},\mathbf{y})\in U_{ad}} J(\mathbf{u}), \tag{1}$$

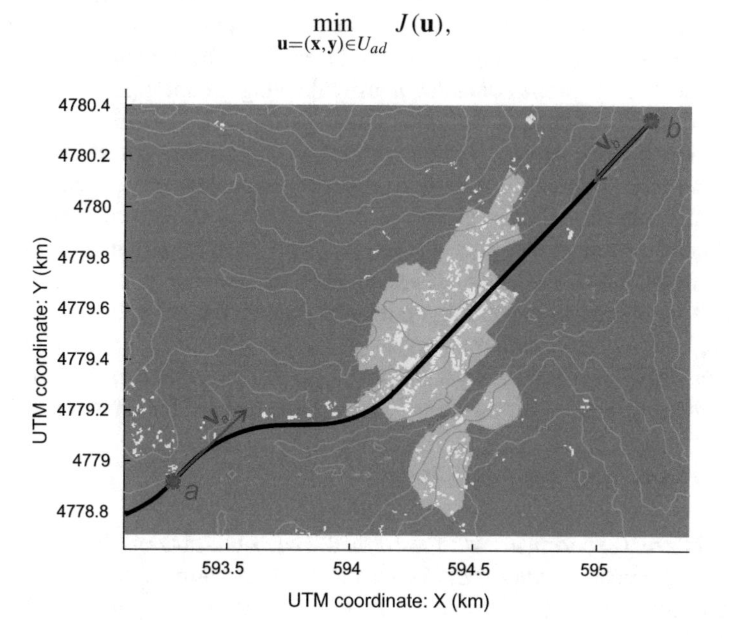

Fig. 1 Map of Parga, NW of Spain (case study), with some necessary information: topography data (contour lines), urban area, buildings and current railway path with points (a and b) where the new bypass has to connect

where **x** and **y** represents the design variables of the horizontal and vertical alignment, respectively, U_{ad} is the admissible set collecting all the constraints that any alignment must fulfil (minimum radii, maximum slopes, etc.), and $J = J_L + J_{LA} + J_{EW}$ represents the final construction cost, where

- J_L stands for the localization costs which include land acquisition and penalties for demolishing buildings or passing through the urban area,
- J_{LA} stands for the length and surface dependent costs, which include platform, ballast, sub-ballast, rails, etc.,
- J_{EW} stands for the earthwork costs.

The first step to address problem (1) is to choose the points where the new bypass must be connecting with the current layout. This is a technical decision that must be taken according to various criteria, considering as first priorities, in this case, the optimal use of the existing infrastructure (shortest bypass length) and, as much as possible, an easy geometrical connection between the two layouts. To simplify the problem, we consider to make this connection firstly in tangents (which decreases the number of design variables of the horizontal alignment [1]), and take advantage of a bridge over the river east of the village, and an overpass in the west, recently built to allow the crossing of a secondary road. The points a and b that were initially chosen to make the connections can be seen in Fig. 1.

The next step is to set the technical constraints allowing us to define the set U_{ad}, and assemble the necessary information to be able to compute the cost of any admissible path. We take a set U_{ad} that guarantees radii, tangents and transition curves bounded below, and slopes bounded above, allowing the minimum and maximum thresholds change in each of the experiments performed. Regarding the evaluation of the costs, besides to consider current values for railway construction, a fixed acquisition price was set outside the forbidden areas (urban area and buildings), cartography data was taken every five metres, and in the absence of geological data, two different materials were considered, homogeneously distributed in two layers of constant thickness (the model allows to take into account various material layers with different thickness depending on the horizontal location).

Finally, for a given number of curves (N) and a given number of slope changes (M), problem (1) is solved using a gradient-type method, which incorporates global optimization techniques for avoiding local minima (see [9] for more details).

3 Numerical Experiments

Several different experiments were performed. The results obtained in two of them are presented below: the first one tries to show the good behaviour of the proposed model, while the second one tries to respond to the problem suggested by the neighbourhood association.

3.1 Experiment 1: Checking the Algorithm

This first example was chosen with the purpose to show the good performance of model (1), trying to replicate an "a priori" known solution. With this aim all buildings were taken as a prohibited area, and the existence of an urban area was neglected. Problem (1) was solved to find a path with $N = 2$ curves and $M = 3$ slope changes joining the initially chosen points. The optimal solution of this problem is the existing railway layout (see Fig. 1). As can be seen in Figs. 2 and 3 the result given by the model is truly satisfactory. Starting from a random initial layout which implies the demolition of numerous buildings (see Fig. 2) and causes a huge earthwork (see Fig. 3), the solution obtained reflects almost with complete accuracy the current railway layout which, as expected, barely causes any earthwork (see Fig. 3).

3.2 Experiment 2: Obtaining a Bypass Surrounding the Urban Area

The second experiment tries to find a bypass fulfilling the requirements of #PARGANONSEDIVIDE movement and, therefore, the set compounding the buildings and the urban zone was taken as a prohibited area. By assuming that $N, M \leq 3$, problem (1) was solved starting from several initial paths, but never an acceptable solution was reached. All the obtained layouts involve the demolition at least one building, the one located north-east of the terminal point a, moving due $\mathbf{v}_a$ which is the direction of connection with the existing railway (see Fig. 1). To tackle this

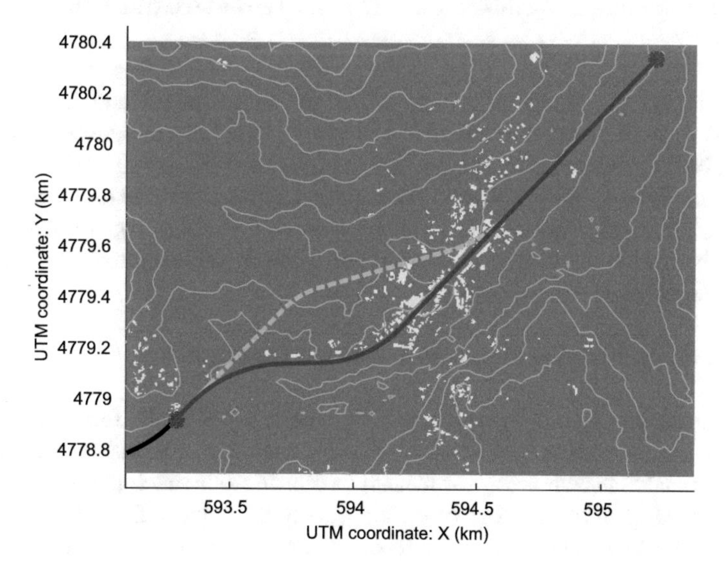

Fig. 2 Experiment 1: Map of Parga with an initial random bypass (dashed) and with the optimal bypass (solid) obtained if the urban area is not taken into account

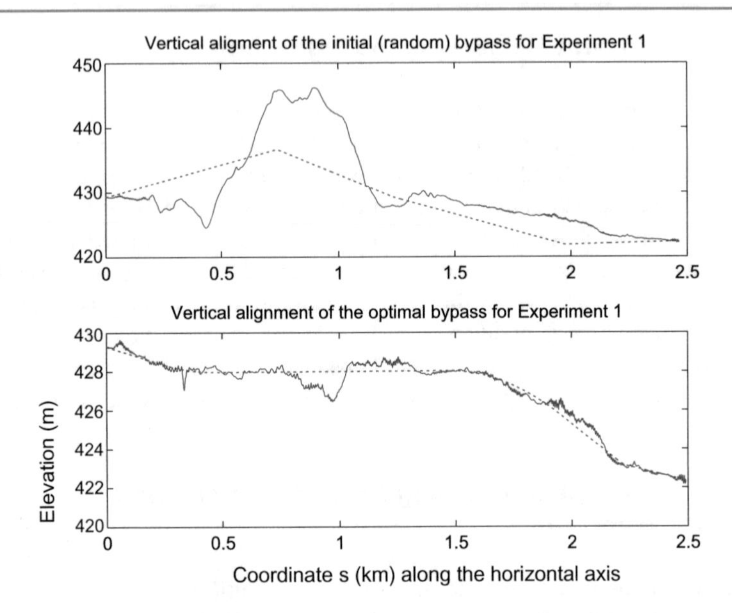

Fig. 3 Experiment 1: Vertical alignment (dashed) and terrain height on the railway central axis (solid), for an initial random bypass (up) and for the optimal bypass obtained if the urban area is not taken into account (down)

difficulty, we proceeded to solve the problem in two stages as follows.

Stage 1: Obtaining an Admissible and Optimal Layout from an Economic Point of View

Firstly, the connection point a was moved 30 m eastwards and, by assuming $N, M \leq 3$, problem (1) was solved to obtain an optimal layout between that translated point, a', and point b. The optimal solution, corresponding with $N = 3$ and $M = 1$, can be seen in Figs. 4 and 5. The optimal bypass skirts the urban area avoiding all buildings, although for accomplish it the new layout has to go through an area with higher terrain elevation, which requires a considerable earthwork in terms of excavation volume.

Stage 2: Connecting the Bypass with the Current Layout

The path obtained in the previous stage links perfectly with the former at point b, but not at the new point a', which is 30 m away towards east. To tackle this issue, the proposal consists in connecting, by means of a transition curve (clothoid), the circular curve west of point a', with one of the tangents of the bypass. This transition curve exists if the tangent and the circumference are close enough and do not intersect. In this case, it can be obtained by the numerical algorithm detailed in the Appendix. This algorithm was used to connect the current layout with the longest tangent of the

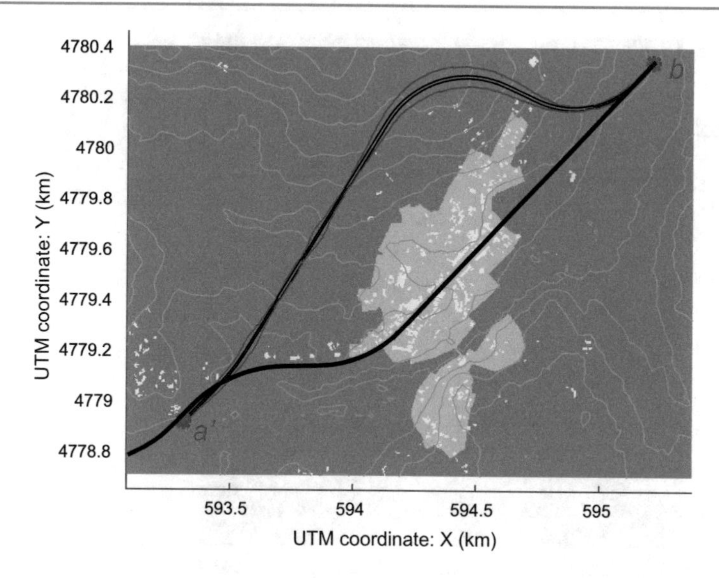

Fig. 4 Experiment 2: Map of Parga with the current railway path (thick solid black) and with the optimal bypass (double thin black) obtained if the west terminal point is displaced 30 m east. Earthwork area is also drawn

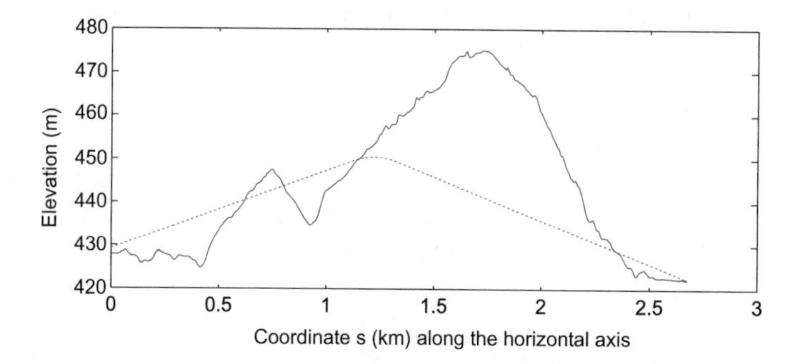

Fig. 5 Experiment 2: Vertical alignment (dashed) and terrain height on the railway central axis (solid), for the optimal bypass obtained if the west terminal point is displaced 30 m east

bypass computed in the previous stage. The final result can be seen in Fig. 6, where the transition curve enabling this connection is zoomed-in.

4 Conclusions and Future Work

In this work, a railway bypass is obtained on the line A Coruña-Palencia (Spain). It skirts the urban area, avoids all buildings, and is optimal from an economic point of view. To obtain that bypass, a model proposed by the authors in a previous work [9] is used. The developed numerical experiments show the good behaviour of the model, but they also bring to light problems that arise when it comes to avoid obstacles,

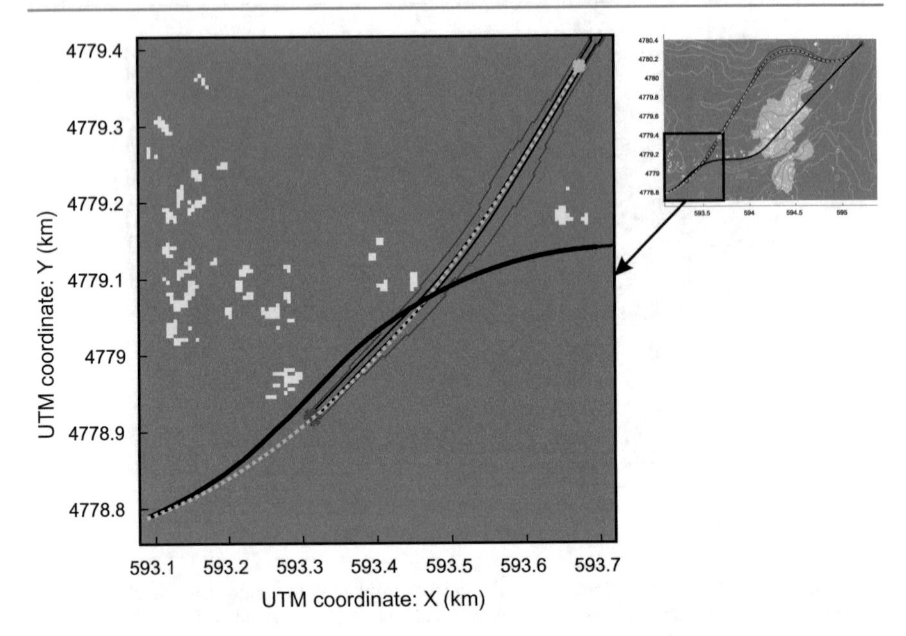

Fig. 6 Map of Parga with the final result (dashed) obtained by a slight modification of the optimal bypass (double thin black) to get a correct connection with the current railway path (thick solid black). Zoom of the modification area

if the connection points are set in advance. Additionally, an algorithm is proposed to link a circular curve with a tangent, which can be very useful in many practical applications. Finally, it should be pointed out that, from what has been observed in this work, for designing railway bypasses using techniques presented in [9] it will be very useful to improve the model in two aspects:

1. Instead of fixing in advance the terminals of the bypass, it will be interesting to fix only the sections (tangent or circular curve) containing them. The specific points should be included among the decision variables of the model.
2. To avoid level crossings, it will be interesting to include in the model the existence of other infrastructures (mainly roads) forcing the new railway to cross them respecting a minimum height between both infrastructures.

These two aspects will be studied in a forthcoming work done by the authors.

5 Appendix: Calculation of the Clothoid Connecting an Oriented Circumference with an Oriented Tangent

Let be r_N an oriented tangent passing through point $P_N = (x_N, y_N)$, and let Φ_N be the angle between r_N^+ and OX^+ (positive semi-axis of abscissas). Let us also consider a circumference C_E of centre $c_E = (c_1, c_2)$, radius R_E (see Fig. 7) which turns in the direction indicated by the parameter $\lambda_E = \pm 1$ ($\lambda_E = -1$ clockwise and $\lambda_E = 1$ counterclockwise).

We analyse, first of all, in what situation the existence of a clothoid arc connecting r_N^+ with C_E is guaranteed. To do so, we denote by $\alpha \in [0, 2\pi)$ and $\beta \in (0, \pi/2)$ the angles formed by the vector $\mathbf{P_N c_E} = (c_1 - x_N, c_2 - y_N)$ with OX^+ and with the tangent to C_E at point P_N, respectively (see Fig. 7). In order to r_N and C_E be connected by a clothoid arc, it is necessary that they do not intersect and have compatible orientations (see [7]). This holds if $\Phi_N \in (\alpha + \beta, \alpha - \beta + \pi)$ and $\lambda_E = -1$, or if $\Phi_N \in (\alpha + \beta - \pi, \alpha - \beta)$ and $\lambda_E = 1$ (see Fig. 7). If the connection has to take place in the r_N^+ ray, it is also necessary a short distance from this ray to the circumference. This is verified if we take $\delta \in (0, \pi - 2\beta)$ small enough, establish

$$\Phi_{min} = \begin{cases} \alpha - \beta - \delta & \text{if } \lambda_E = 1, \\ \alpha + \beta & \text{if } \lambda_E = -1, \end{cases}$$

$$\Phi_{max} = \begin{cases} \alpha - \beta & \text{if } \lambda_E = 1, \\ \alpha + \beta + \delta & \text{if } \lambda_E = -1, \end{cases}$$

and guarantee that

$$\Phi_{min} < \Phi_N < \Phi_{max}. \tag{2}$$

Let us suppose now that Φ_N fulfils (2). Let us show how to compute the clothoid arc connecting r_N with C_E. That arc will cover from curvature 0 (tangent r_N) to

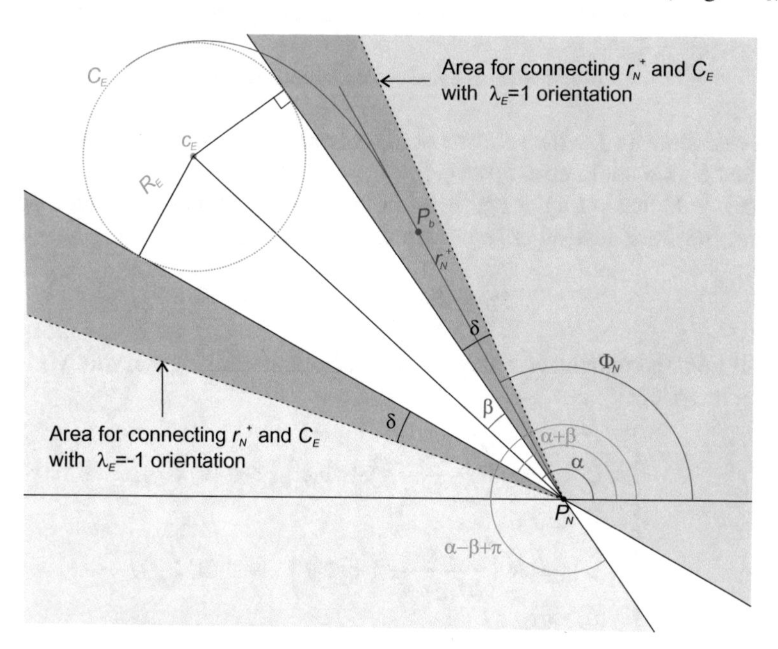

Fig. 7 Notation and areas where it is possible to connect an oriented circumference C_E with an oriented straight segment r_N

curvature $1/R_E$ (circumference C_E), and it must do it at a constant ratio, therefore if its measure is $L > 0$ m, the scale parameter should be $|\nu_c| = 1/(LR_E)$. Thus, for every $L > 0$, the clothoid arc starting at point P_N, tangent to r_N, with curvature 0, direction given by λ_E, and that at L m it has curvature $1/R_E$ is given by the solution of the following system (see [10]):

$$\begin{cases} x'(s) = \cos\left(\frac{\lambda_E}{2LR_E}s^2 + \Phi_N\right), \; s \in (0, L), \\ x(0) = x_N, \\ y'(s) = \sin\left(\frac{\lambda_E}{2LR_E}s^2 + \Phi_N\right), \; s \in (0, L), \\ y(0) = y_N. \end{cases} \tag{3}$$

It is easy to see that the circumference connecting this clothoid arc at point $(x(L), y(L))$ and radius R_E is the one centred at

$$c(L) = (x(L), y(L)) + \lambda_E R_E(-\sin(\Phi(L)), \cos(\Phi(L))),$$

being $\Phi(L) = \frac{\lambda_E L}{2R_E} + \Phi_N$.

We want that circumference to be C_E and therefore:

1. The distance between c_E and r_N has to be the same as the distance between $c(L)$ and r_N, i.e., L has to be the solution of the following non linear equation

$$\text{distance}(c(L), r_N) - \text{distance}(c_E, r_N) = 0. \tag{4}$$

   Let us denote by L_E the solution of (4), whose existence is guaranteed because r_N and C_E do not intersect (see [7]).
2. It must hold that $c(L_E) = c_E$, therefore is necessary to move the initial point of the clothoid arc. Instead of beginning at P_N, it must start at point

$$P_b = P_N + (c_E - c(L_E)). \tag{5}$$

With all this, the parameterization of the clothoid arc linking r_N with C_E, is the solution of system

$$\begin{cases} x'(s) = \cos\left(\frac{\lambda_E}{2L_E R_E}s^2 + \Phi_N\right), \; s \in (0, L_E), \\ x(0) = x_b, \\ y'(s) = \sin\left(\frac{\lambda_E}{2L_E R_E}s^2 + \Phi_N\right), \; s \in (0, L_E), \\ y(0) = y_b, \end{cases} \tag{6}$$

being $L_E > 0$ the solution of (4) and $(x_b, y_b) = P_b$ given by (5).

The solution of (6) can be obtained by using any numerical method for solving ordinary differential equations (ODEs) [8]. However, it should be noted that before solving it, it is necessary to obtain the value of L_E. If an iterative method is used to solve (4) and compute L_E, the solution of (3) corresponding to L_E is obtained, and it is no longer necessary to solve (6). The solution of (6) can be computed by adding $c_E - c(L_E)$ to the solution of (3) corresponding to L_E. Anyway, in every iteration of the method used to solve (4), it would be necessary to solve system (3), and at this step a numerical method of ODEs has to be applied.

References

1. Casal, G., Santamarina, D., Vázquez-Méndez, M.E.: Optimization of horizontal alignment geometry in road design and reconstruction. Transpot. Res. C-Emer. **74**, 261–274 (2017)
2. Federación castellano manchega de amigos del ferrocarril: Planes de supresión y seguridad en pasos a nivel (in spanish). http://www.fcmaf.es/PEIT/PEIT_al_dia/PS_PNivel.htm#PlanSupresion. Cited 16 Sep 2019
3. Hirpa, D., Hare, W., Lucet, Y., Pushak, Y., Tesfamariam, S.: A bi-objective optimization framework for three-dimensional road alignment design. Transpot. Res. C-Emer. **65**, 61–78 (2016)
4. Li, W., Pu, H., Zhao, H., Hu, J., Meng, C.: Intelligent railway alignment optimization based on stepwise encoding genetic algorithm. J. Southwest Jiaotong Univ. **48**(5), 831–838 (2013)
5. Mondal, S., Lucet, Y., Hare, W.: Optimizing horizontal alignment of roads in a specific corridor. Comput. Oper. Res. **64**, 130–138 (2015)
6. Pu, H., Zhang, H., Li, W., Xiong, J., Hu, J., Wang, J.: Concurrent optimization of mountain railway alignment and station locations using a distance transform algorithm. Comput. Ind. Eng. **127**, 1297–1314 (2018)
7. Stoer, J.: Curve fitting with clothoidal splines. J. Res. Nat. Bur. Stand. **87**(4), 317–346 (1982)
8. Vázquez-Méndez, M.E., Casal, G.: The clothoid computation: a simple and efficient numerical algorithm. J. Surv. Eng. **142**(3), 04016005 (2016)
9. Vázquez-Méndez, M.E., Casal, G., Castro, A., Santamarina, D.: A 3D model for optimizing infrastructure costs in road design. Comput.-Aided Civ. Inf. **33**, 423–439 (2018)
10. Vázquez-Méndez, M.E., Casal, G., Ferreiro, J.B.: Numerical computation of egg and double-egg curves with clothoids. J. Surv. Eng. **146**(1), 04019021 (2020)

Reduced Models for Liquid Food Packaging Systems

Nicola Parolini, Chiara Riccobene and Elisa Schenone

Abstract

Simulation tools are nowadays key elements for effective production, design and maintenance processes in various industrial applications. Thanks to the advances that have been achieved in the past three decades, accurate and efficient solvers for computational fluid dynamics and computational mechanics are routinely adopted for the design of many products and systems. However, the most accurate models accounting for the complete three-dimensional complex physics (of even multi-physics) are not always the best option to pursue, in particular in the preliminary design phase or whenever very fast evaluations are required. In this paper, we present a set of reduced numerical models that have been developed in the past few years to support the design of paperboard packaging systems.

1 Introduction

Food packaging is an important industrial sector in which the development of highly specialized automation systems plays a fundamental role. The entire process should be able to meet the tight standards for safety and quality control in food industry, as well as the demand of high reliability and productivity in complex industrial plants.

N. Parolini (✉)
MOX, Dipartimento di Matematica, Politecnico di Milano, P.zza Leonardo da Vinci 32, 20133 Milan, Italy
e-mail: nicola.parolini@polimi.it

C. Riccobene · E. Schenone
MOXOFF s.p.a., via Schiaffino 11/19, 20159 Milan, Italy
e-mail: chiara.riccobene@moxoff.com

E. Schenone
e-mail: elisa.schenone@moxoff.com

P. Quintela Estévez et al. (eds.), *Advances on Links Between Mathematics and Industry*, SxI - Springer for Innovation / SxI - Springer per l'Innovazione 15,
https://doi.org/10.1007/978-3-030-59223-3_3

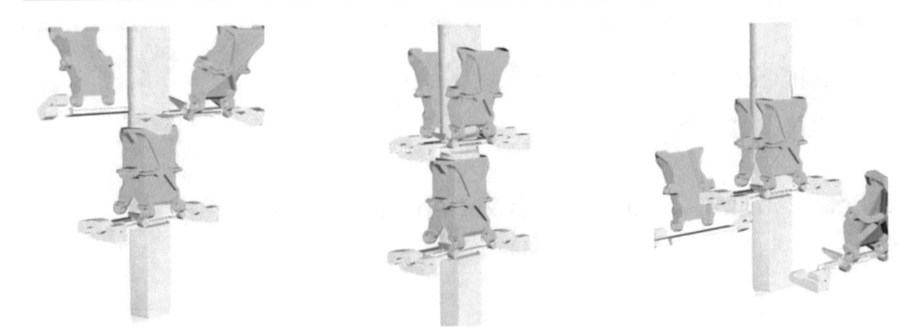

Fig. 1 The action of the jaw system that shapes a cylindrical carton tube in boxes (courtesy of Tetra Pak Packaging Solution s.p.a.)

Laminated paperboard is a widely used material in the packaging industry, since box-shaped solids can be easily obtained from flat paper roll, and the formed packages are light and stiff. The most efficient way to assemble and fill carton-based boxes consists in deforming a cylindrical sleeve filled with liquid, using a jaw system in which mechanical clamps enter periodically in contact with the carton tube squeezing it until it is closed and shaping its lateral sides to a rectangular section (see Fig. 1).

In this paper, we will consider a specific packaging technology, for the assembling of carton-based liquid food packages in highly performing automatized filling machines, able to assemble several packages per seconds in an aseptic environment with a fully automatic material supply.

2 A Complex FSI Problem

A filling machine is a complex system in which, starting from a roll of laminated paperboard and a continuous liquid food supply, filled bricks are assembled at a production rate of several thousands per hour. The paperboard is initially unrolled and is subject to a sterilization process crossing a peroxide bath, then is wrapped and vertically sealed in a cylindrical tube moving downward around the injector pipe. At the end of the process, which occurs around the outflow of the injector pipe, the bricks are assembled through the following steps (sketched in Fig. 2):

(a) the bottom end of the package is folded and sealed;
(b) the package is laterally shaped by the jaw system;
(c) the top end of the package is closed and the bottom end is cut;
(d) the semi-finished package is released.

The fluid is injected into the carton tube through the co-axial injector pipe. The carton tube is deformed by the combined action of the jaw system and the action of the fluid, whose motion is, in turn, affected by the carton tube deformation. This coupled interaction defines a complex fluid-structure interaction (FSI) problem.

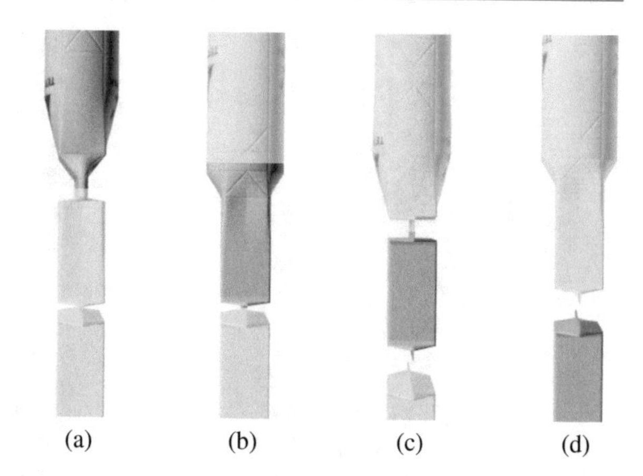

Fig. 2 Different phases of the forming cycle: **(a)** bottom closure **(b)** side forming **(c)** top closure **(d)** package release

A simulation of the complete three-dimensional FSI problem coupling the incompressible Navier-Stokes equations governing the product fluid dynamics with the elasticity equations governing the structural deformation has been subject of research for the specific application at hand (see, e.g., [1,20,21]), as well as applications which share some key characteristics of the considered problem, as the strongly coupled nature of the FSI coupling [2,5–8,16,17]. In particular, it is well known from the literature that most staggered fluid-structure interaction coupling can be subjected to numerical instabilities whenever the so-called *added mass effect* is present [3], which is most often the case when a liquid interacts with a deformable thin structure. In such conditions, a strongly coupled FSI algorithm is necessary, which typically requires a number of sub-iterations at each time step solving for the fluid and the structural problems. As a consequence, the computational cost of a full 3D FSI simulation for this kind of applications is most of the time unaffordable and resorting to reducing strategies becomes unavoidable.

3 Geometrical Model Reduction

A first numerical reduction approach for the filling machine has been obtained deriving a geometrical reduced mathematical model governing the fluid-structure interaction between the filling liquid and the papertube. The filling portion of the packaging system has been modeled adopting a low dimensional (1D-0D) geometrical multiscale model inspired by a similar simulation framework that has been successfully adopted in the past few years for modeling the human cardiovascular system [9–11,22].

This kind of models can be extremely useful in the design process of the packaging machine as they allow fast (in some cases, real time) simulations able to capture the main flow features associated to the pulsatile flow generated by the periodic contraction of the papertube. In particular, they are used to simulate the propagation of the strong pressure waves that are produced by the periodic squeezing of the

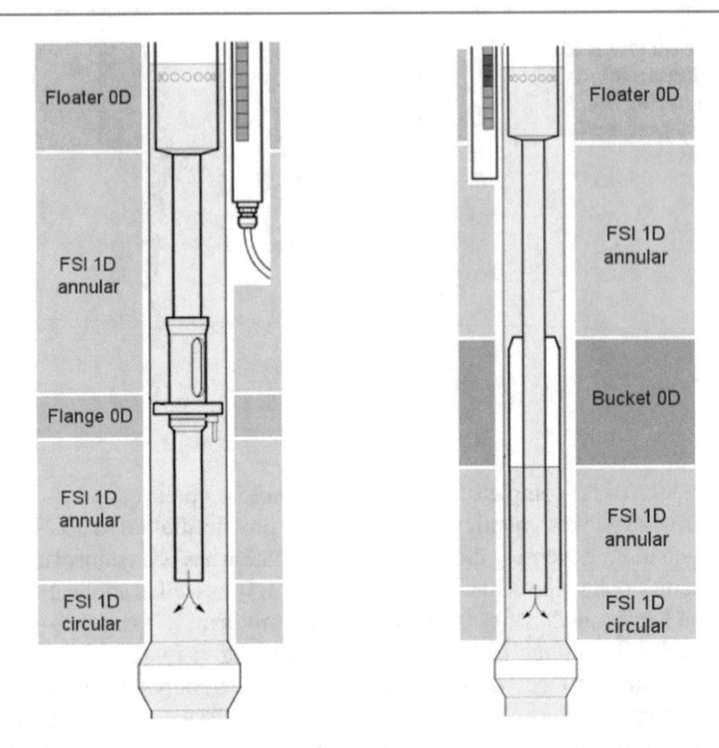

Fig. 3 Two pressure wave dampener technologies: counter-pressure flange (left) and air bucket (right)

papertube by the jaw system. The objective of this kind of analyses is the design of effective tools to limit their impact on the final quality of the packages.

In this regard, two are the main technologies that are usually adopted in packaging systems: the first in which the pressure waves are dampened constraining the pulsatile flow to move across small orifices distributed on a *counter-pressure flange*; the second where the dampening effect is obtained using an *air bucket* which is periodically filled and emptied. Two sketches of the packaging systems each equipped with one of the pressure dampeners are displayed in Fig. 3. For both configurations, a constant inflow is supplied by the injector, while a time dependent flowrate associated to the squeezing action of the jaw system can be prescribed at the lower boundary. Indeed, in the reduced model, we will prescribe at lower boundary the net flowrate given by the difference between the constant flowrate through the injector and the time-dependent flowrate associated to the change of volume imposed by the jaw system (that has been estimated solving the full 3D package deformation using the software Abaqus).

A set of geometrical reduced models, one for each of the sub-domains has been identified, together with suitable coupling and boundary conditions.

In order to model the machine equipped with the flange (Fig. 3, left), we need to define a 1D model governing the flow in the cylindrical domain under the injector, a similar 1D model for the flow in the annular domains around the injector and 0D models accounting for the flange and the floater. These models are accommodated in series and the coupling conditions impose the continuity of flowrate and pressure at the interface between different models.

A similar model coupling 1D and 0D elements has been developed for the machine equipped with the air bucket (Fig. 3, right). Also in this case, 1D FSI models for circular and annular sections are required, while the air bucket and the floater are modeled by 0D models. In this case, the flowrate imposed on the bottom surface is split between a portion that enters the bucket (changing the liquid level inside it and, therefore, the pressure of the enclosed air) and the remaining part that flows through the thin annular interspace between the bucket and the papertube. Thus, in this case, the coupling condition should also account for this flow bifurcation by prescribing a balance of flowrates and the continuity of pressure at the bifurcation point.

Note that, in order to impose the coupling conditions at each time step of the numerical solution, the different models may be solved iteratively until continuity of pressure and flowrate is achieved at each interface (see [19]).

For both configurations, the flowrate is imposed on the bottom boundary while a pressure boundary condition (accounting for the hydrostatic level) is imposed on the upper boundary. As will be discussed in Sect. 3.1, this choice of boundary conditions is consistent with the hyperbolic nature of the 1D models.

Hereafter, we briefly describe the different 1D and 0D models that are used for the two types of systems (flange and bucket), starting from the mono-dimensional FSI models that govern the flow within the deformable papertube.

3.1 1D FSI Model

Mono-dimensional FSI models for incompressible flows interacting with elastic vessels have been developed and extensively used for the simulation of the human cardiovascular system (see, e.g., [9, 11, 22]). These models work under the assumptions of axial symmetry, small displacement of the structure (usually limited to normal direction), dominance of the axial component. Under these hypotheses, by integrating the incompressible Navier-Stokes equations over the cross section and introducing a simplified (algebraic) model for the elastic response of the structure, a system of partial differential equations can be derived for $z \in \Omega \subset \mathbb{R}$, $t > 0$, namely,

$$\frac{\partial A}{\partial t} + \frac{\partial Q}{\partial z} = 0, \tag{1a}$$

$$\frac{\partial Q}{\partial t} + \frac{\partial}{\partial z}\left(\alpha \frac{Q^2}{A}\right) + \frac{A}{\rho}\frac{\partial p}{\partial z} + K_r\left(\alpha \frac{Q}{A}\right) + Ag = 0, \tag{1b}$$

where $A(z,t)$ is the section area at longitudinal position z at time t, $Q(z,t)$ is the flowrate, $p(z,t)$ is the mean pressure on the section, α is the Coriolis coefficient (which depends on the velocity profile over the section), K_r is a friction coefficient and g is a specific body force (most often the gravity acceleration). In the simplest case, for cylindrical domains, the structural response can be described by an algebraic relation between the pressure difference across the structure and the normal displacement η,

$$p - p_{\text{ext}} = b\eta,$$

where p_{ext} is the prescribed pressure distribution on the external side of the structure and the stiffness parameter b is defined as

$$b = \frac{Eh}{(1 - \xi^2)R_0^2}, \tag{2}$$

with E and ξ denoting the Young modulus and the Poisson ratio of the material, respectively, h its thickness and R_0 is the radius of the reference (initial) configuration.

In order to close system (1), the structural response can be more conveniently formulated as an algebraic relation between the flow pressure p and the area A given by

$$p - p_{\text{ext}} = \psi(A; A_0, \beta);$$

where $A_0 = \pi R_0^2$ and β is a stiffness parameter. The exact definition of ψ and β will be detailed in the following for the different domain geometries (circular or annular).

System (1) is an hyperbolic system for the conservative variables (A, Q) that is here numerically solved using a finite-element Taylor-Galerkin method [9]. The characteristic analysis of system (1), that has been carried out in [22], shows that, for the case at hand, the Jacobian matrix has one positive eigenvalue, corresponding to a characteristic wave moving in positive z direction and a negative one, corresponding to a second characteristic wave moving backwards. Therefore, one boundary condition has to be imposed on each boundary of the 1D domain.

3.1.1 Closure for Circular Sections

For circular sections, the normal displacement is defined as $\eta = R - R_0 = \frac{\sqrt{A}-\sqrt{A_0}}{\sqrt{\pi}}$, so that the closure of the 1D model (1) can be obtained (as in [11]), by choosing

$$\psi(A; A_0, \beta) = \beta(\sqrt{A} - \sqrt{A_0}), \tag{3}$$

where the stiffness coefficient is given by

$$\beta = \frac{\sqrt{\pi}Eh}{(1 - \xi^2)A_0}. \tag{4}$$

The Coriolis coefficient α is defined as follows

$$\alpha = \frac{\int_S s^2 d\sigma}{A},$$

where we consider an axisymmetric velocity profile $s(r)$ over the section S as the power law radial function $s(r) = \gamma^{-1}(\gamma + 2)(1 - (r/R)^\gamma)$. The friction coefficient can be defined as

$$K_r = -2\pi \nu s'(1) = 2\pi \nu(\gamma + 2),$$

where ν is the kinematic viscosity of the fluid,

For $\gamma = 2$, we have a parabolic profile which gives $\alpha = 4/3$ and $K_r = 8\pi\nu$, while for the typical choice $\gamma = 9$, we get $\alpha = 1.1$ and $K_r = 22\pi\nu$.

3.1.2 Closure for Annular Domain

In order to consider the annular sub-domains around the injector and the paper tube, a similar 1D model has been developed. Here the area of the annular section is given by $A = \pi(R^2 - R_i^2)$, where R_i denotes the internal radius on the annulus. Similarly, the area of the reference section is $A_0 = \pi(R_0^2 - R_i^2)$.

To close system (1) we define the normal displacement as $\eta = R - R_0 = \frac{\sqrt{A+A_i} - \sqrt{A_0+A_i}}{\sqrt{\pi}}$, where $A_i = \pi R_i^2$, and we obtain

$$\psi(A; A_0, \beta) = \beta(\sqrt{A + A_i} - \sqrt{A_0 + A_i}) \tag{5}$$

with β defined as in (4).

In this case the Coriolis and friction coefficients are computed assuming an axisymmetric parabolic profile on the annular section, namely, $s(r) = ar^2 + br + c$ with $a = -\frac{4}{(R_0 - R_i)^2}$, $b = -(R_0 + R_i)a$ and $c = -aR_i^2 - bR_i$.

3.2 0D Model of the Counter-Pressure Flange

The contribution of the counter-pressure flange is accounted for with a 0D lumped model, prescribing a concentrated pressure loss as a function of the flowrate. The flange works as a pressure wave dampener thanks to the dissipation occurring on the flow which is confined in a number of small orifices distributed on the flange surface. The orifices are usually designed such that the loss coefficient for the flow moving upwards is larger than the one for the recovery phase when the flow is moving downward across the flange. Assuming N identical orifices on the flange, the 0D flange model is given by

$$\Delta p = p_{\text{under}} - p_{\text{over}} = \frac{1}{2}\rho\,\text{sign}(Q)\left(\frac{Q}{N A_i}\right)^2 C_d, \tag{6}$$

where Δp is the pressure difference across the flange (positive for upwards flowrate), A_i is the area of single orifice and C_d is the pressure loss coefficient (that needs to be estimated).

3.3 0D Model of the Air Bucket

Air buckets are typically used in packaging system with higher production rates. In this case the dampening effect is given by the combined action of the energy absorbed by the gas inside the bucket which is periodically compressed and the viscous dissipation occurring on the liquid flowing in the tiny annular domain between the bucket and the papertube (see Fig. 3).

A 0D model computing the gas pressure inside the bucket is obtained considering the ideal gas law ($pV = \text{const}$) and computing the gas volume by updating the liquid level inside the bucket as a function of the flowrate entering the bucket, namely,

$$h_l(t) = \left(h_0 + \frac{1}{A_b} \int_0^t Q_b \, d\tau \right), \tag{7}$$

where h_0 is the initial level of liquid in the bucket, A_b is the area of the section of the bucket and Q_b is the flowrate entering the bucket. Indicating with L the length of the bucket, the volume of the gas inside the bucket is $V(t) = A_b(L - h_l(t))$.

An additional 0D model accounting for the floater can be added to the system. As sketched in Fig. 3, the floater is a light co-axial cylinder placed above the pressure dampener and moving vertically along the injector, used to measure the level of liquid inside the papertube during the different phases of the filling cycle. Even if ideally the floater should move with the free-surface level, due to the strong inertia associated to the fast level variations, a model able to capture the dynamics of the floater is required. Indeed, the dynamics of the floater seems to play an additional (non-negligible) role in the pressure wave dampening and including it in the model can be useful to take into account these effects.

3.4 Numerical Results of the Geometrical Reduced FSI Models

For the problem at hand, some of the hypotheses that have been introduced to derive the geometrical reduced models are not completely fulfilled (for instance, close to the injector the transverse velocity components may not be negligible). However, the reduce models introduced in the previous section have proved to be able to capture the main flow features of the system, in particular concerning the pressure wave propagation.

These models have been adopted to simulate a number of different packaging machines of both types. In Fig. 4 (left), we compare, for a machine with the counter-pressure flange, the numerical prediction and the experimental measurements of the time evolution over two periods of the pressure level under the flange. We can observe that amplitude and phase of the two main pressure peaks are well approximated. Notice that some local maxima included in the experimental pressure signal are not detected by the numerical simulation. These features are associated to specific process phases that could not be included in the considered geometrically reduced 1D model, such as, for instance, the shape change of the papertube section during

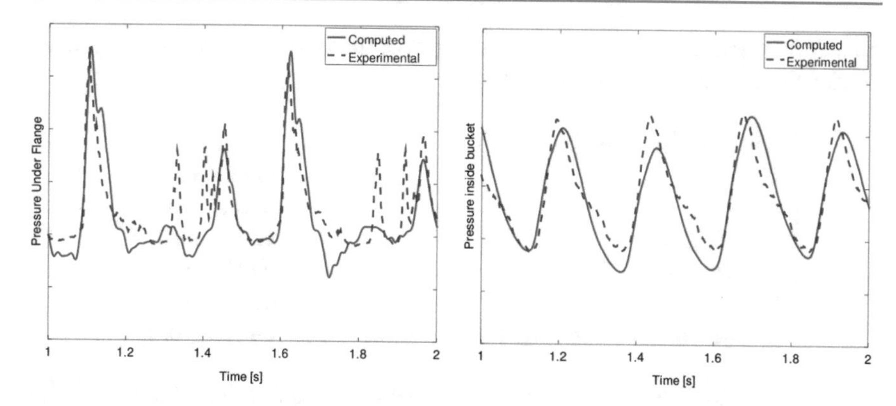

Fig. 4 Comparison with experimental measurements for the pressure drop across the flange (left) and the gas pressure inside the bucket (right)

the jaw action from a circular to a rectangular shape. For confidentiality limitations, the pressure scale in Fig. 4 has been removed.

Similar results, also showing an good matching between the numerical results and the experimental measurements, have been obtained for air bucket systems. In Fig. 4 (right), the comparison related to the gas pressure inside the bucket is displayed.

4 Reduced 3D FSI Using Transpiration and Absorbing BCs

As we have seen, the geometrical reduction represents a possible reduction strategy allowing to capture some relevant characteristics of the coupled system. However, some aspects of the problem can only be captured resorting to the simulation of the complete flow field in 3D.

Since we have seen that the solution of a complete 3D FSI problem may be prohibitively expensive due to the need of a strongly coupled FSI approach, it is worth trying to devise a simulation framework for FSI simulation with an intermediate level of complexity. In this respect, the interaction between the fluid and the deformable walls may be modeled by the so-called *transpiration boundary conditions*. As in [4, 12], for small displacements, the wall structural response can be modeled as an elastic shell where the tangential components of the wall stress and of the displacement are neglected. The main advantage of this approach, that has been extensively exploited in aeronautical applications to solve aero-elastic problems, is that it does not require to change the domain (and thus to move the computational grid) at each time step and the simplified structural model reduces to a boundary condition for the flow problem.

Assuming that the displacement η is only in the normal direction, following the derivation proposed in [4], the transpiration condition can be derived from an elasto-dynamic equation for the normal displacement η defined on the middle surface Σ of the deformable shell. The transpiration condition can be formulated as a Robin-type boundary condition for the normal component of the velocity in the time-dependent Navier-Stokes equations, as follows:

$$\begin{cases} \partial_t \mathbf{u} - \nu \nabla^2 \mathbf{u} + (\mathbf{u} \cdot \nabla)\mathbf{u} + \nabla p = \mathbf{f}, & \text{in } \Omega \times (0, T), \\ \nabla \cdot \mathbf{u} = 0, & \text{in } \Omega \times (0, T), \\ \mathbf{u} = \mathbf{u}_0, & \text{in } \Omega, t = 0, \\ \mathbf{u} = \mathbf{u}_{\text{in}}, & \text{on } \Gamma_{\text{in}} \times (0, T), \\ -\nu \partial_{\mathbf{n}} \mathbf{u} + p\mathbf{n} = 0, & \text{on } \Gamma_{\text{out}} \times (0, T), \\ \mathbf{u} \times \mathbf{n} = 0, & \text{on } \Sigma \times (0, T), \\ \rho_s h \partial_t \mathbf{u} \cdot \mathbf{n} - \nabla_c \cdot (\mathbf{T} \nabla_c \eta) + a\mathbf{u} \cdot \mathbf{n} + b\eta = p, & \text{on } \Sigma \times (0, T), \end{cases} \tag{8}$$

where $\mathbf{u}$ is the fluid velocity, p is the fluid pressure (rescaled with the fluid density), ν is the fluid kinematic viscosity, ρ_s is the structure density, h is the structure thickness, ∇_c is the covariant gradient, T is the stress (possibly prestressed) tensor, a and b are damping and elastic coefficients, respectively (see [4] for details).

The normal displacement η can be computed integrating over time the normal velocity, as

$$\eta := \int_0^t \mathbf{u} \cdot \mathbf{n} ds.$$

The coefficient b depends on the material properties of the deforming structure and characterize its elastic response: the greater b is, the more rigid the structure. In particular, for cylindrical structures, b can be defined as in Eq. (2).

Further simplification of the structural model may be considered when projection-based flow solvers are used, resulting on a pressure boundary condition on the moving wall, as discussed in [12].

A numerical solution of Problem (8) has been obtained using $\mathbb{P}_2/\mathbb{P}_1$ finite element for velocity and pressure and using an implicit Euler time advancing scheme with a semi-implicit treatment of the convective term.

The ability of the model in capturing the propagation of the pressure wave generated by the time-dependent inlet velocity profile $\mathbf{u}_{\text{in}}(t)$ has been investigated. In order to avoid (or at least minimize) the numerical wave reflection that are generated when homogeneous Neumann condition are imposed on the outflow boundary, the linearized absorbing condition (LAC) proposed in [14] is imposed at the outflow boundary. This condition prescribes a linear dependence of the pressure on the flow rate Q:

$$p = \frac{\pi \sqrt{b}}{\sqrt{2\rho} A^{5/4}} Q \qquad \text{on } \Gamma_{\text{out}}, \tag{9}$$

where A is the area of the outflow boundary.

A simple test case to assess the reduced 3D FSI model with transpiration and absorbing boundary conditions has been setup. We consider the propagation of a pressure wave in the flange system. We start from the fluid at rest, an hydrostatic pressure distribution and null flowrate imposed on the bottom boundary. At a given time $t = \bar{t}$, a sinusoidal pulse on the inlet flowrate is started, which causes a pressure

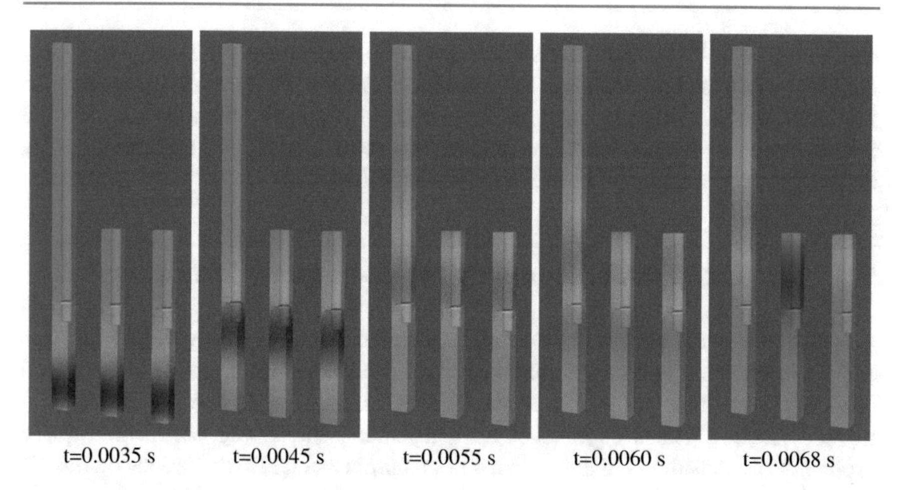

Fig. 5 Propagation of the pressure wave at different time instants. For each time, the solution obtained on a longer domain (left), with homogeneous Neumann outflow condition (middle) and with the linear absorbing condition (right) are reported

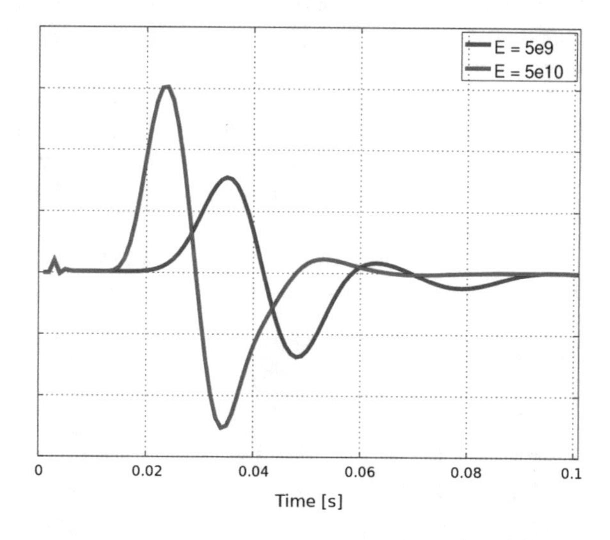

Fig. 6 Time history of the pressure drop on the flange for different values of Young modulus

wave to propagate along the papertube domain, thanks to the deformability of the wall modeled by the transpiration conditions.

In Fig. 5, the solution obtained on a longer domain is compared with two solutions on a shorter domain with and without absorbing boundary conditions. The effectiveness of the linearized absorbing condition in reducing the spurious reflection at the outflow boundary can be clearly appreciated.

A second test case is presented in order to highlight the importance of accounting for FSI effects in this kind of problems, even if with simplified approach as the one based on transpiration boundary conditions. In Fig. 6, we compare the different behaviors of the pressure for different values of the wall rigidities ($E = 5 \times 10^9$ Pa and $E = 5 \times 10^{10}$ Pa). The peaks on the pressure drop across the flange are almost doubling when the higher stiffness is considered and the pressure wave travels much faster. With this results in mind, it is clear that no meaningful indication on the pressure can be obtained when simulating the problem with rigid CFD simulations that do not account, at any level of details, for FSI effects.

5 Reduced-Order Modeling for Multi-query Problems

A further possible reduction approach that is currently being investigated for the considered application is based on reduced-order models (ROM), such as the reduced basis (RB) method [13,18,23] or the Proper Orthogonal Decomposition (POD) method [15,24]. These approaches are particularly interesting when multi-query problems are considered, that is when a large number of CFD evaluations on the 3D geometry are required (such as, for instance, parametric, optimization or uncertainty quantification studies). In all reduced-order models, the approximate solutions can be computed very efficiently (sometime in real time) as a suitable (linear or nonlinear) combination of ad hoc basis functions. The latter are pre-computed in a (computationally expensive) offline phase by solving the high fidelity problem for a limited set of parameters.

In the application at hand, preliminary results have been obtained on the fluid-dynamic analysis of the counter-pressure flange. This kind of analyses can be used, for instance, to obtain the pressure/flowrate characterization for various design parameters in order to evaluate the pressure drop coefficient to be used in the 0D model described in Sect. 3.2.

A parametric study considering two parameters (fluid viscosity and papertube diameter) has been carried on using a reduced basis method and taking as basis

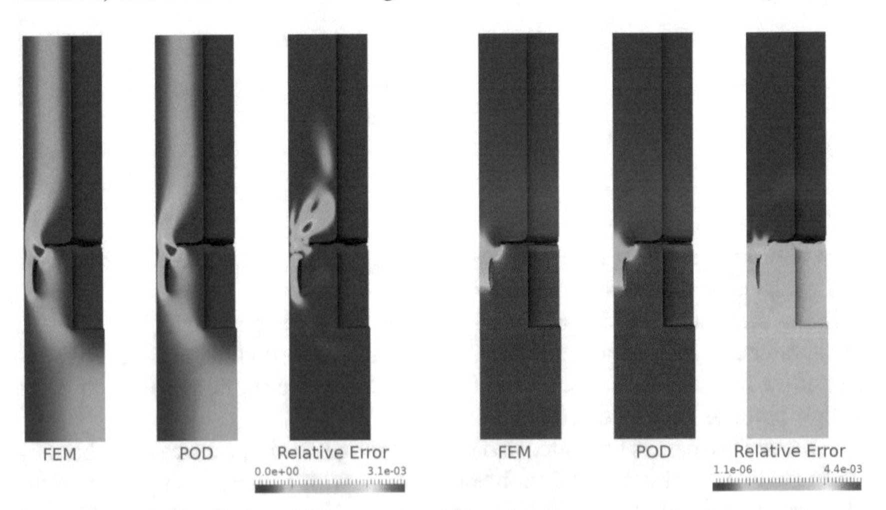

Fig. 7 Comparison of the velocity (left) and pressure (right) fields: for each field the high fidelity FEM solution, the POD solution and the relative error are reported (from left to right)

functions the first 10 modes obtained by a POD computed on 24 snapshots uniformly distributed in the parameter space. In Fig. 7, we report a comparison between the velocity and pressure field computed using a high fidelity finite-element solver and the corresponding reduced solution. The relative error for the solution obtained with a choice of parameters different from those used for the snapshots shows that the reduced solution is able to guarantee a good accuracy. The advantage in terms of computational cost can be appreciated comparing the CPU time required for the high fidelity finite-element solution (4.5 h) with that of the online phase of the reduced method (7.3 min).

6 Concluding Remarks

We have presented a set of numerical simulation tools that have been developed to support the design of packaging systems for liquid food products. Different reduction strategies, ranging from geometrical reduction, to a simplified treatment of 3D FSI problems based on transpiration wall conditions and including reduced-order models for multi-query problems, have been developed and effectively integrated in a complex design framework, in which the role of numerical simulations, at all levels of computational complexities, plays an increasingly important role.

Acknowledgements This research has been supported by Tetra Pak Packaging Solution s.p.a. in the framework of a consolidated research partnership with MOXOFF s.p.a and the MOX Laboratory. The authors are grateful to all the members of Tetra Pak technical team, leaded by Dr. Roberto Borsari, for the insightful and stimulating discussions. Marco Pischedda, Andrea Mola and Jacopo Corno are also acknowledged for their contribution on the development of the tools presented in this work.

References

1. Andersson, L.: Numerical investigations of a partitioned FSI algorithm for Tetra Pak's filling tube. Master's thesis, Lund University (2016)
2. Badia, S., Quaini, A., Quarteroni, A.: Modular vs. non-modular preconditioners for fluid-structure systems with large added-mass effect. Comput. Methods Appl. Mech. Eng. **197**(49-50), 4216–4232 (2008)
3. Causin, P., Gerbeau, J.-F., Nobile, F.: Added-mass effect in the design of partitioned algorithms for fluid-structure problems. Comput. Methods Appl. Mech. Eng. **194**(42–44), 4506–4527 (2005)
4. Chacón Rebollo, T., Girault, V., Murat, F., Pironneau, O.: Analysis of a coupled fluid-structure model with applications to hemodynamics. SIAM J. Numer. Anal. **54**(2), 994–1019 (2016)
5. Colciago, C., Deparis, S., Quarteroni, A.: Comparisons between reduced order models and full 3D models for fluid-structure interaction problems in haemodynamics. J. Comput. Appl. Math. **265**, 120–138 (2014)
6. Crosetto, P., Deparis, S., Fourestey, G., Quarteroni, A.: Parallel algorithms for fluid-structure interaction problems in Haemodynamics. SIAM J. Sci. Comput. **33**(4), 1598–1622 (2011)
7. Deparis, S., Discacciati, M., Fourestey, G., Quarteroni, A.: Fluid-structure algorithms based on Steklov-Poincaré operators. Comput. Methods Appl. Mech. Eng. **195**(41–43), 5797–5812 (2006)

8. Formaggia, L., Gerbeau, J.F., Nobile, F., Quarteroni, A.: On the coupling of 3d and 1d Navier-Stokes equations for flow problems in compliant vessels. Comput. Methods Appl. Mechan. Eng. **191**(6–7), 561–582 (2001)
9. Formaggia, L., Lamponi, D., Quarteroni, A.: One-dimensional models for blood flow in arteries. J. Eng. Math. **47**(3), 251–276 (2003)
10. Formaggia, L., Moura, A., Nobile, F.: On the stability of the coupling of 3d and 1d fluid-structure interaction models for blood flow simulations. ESAIM Math. Modell. Numer. Anal. **41**(4), 743–769 (2007)
11. Formaggia, L., Quarteroni, Q., Veneziani, A.: Cardiovascular Mathematics: Modeling and Simulation of the Circulatory System. Springer (2009)
12. Gostaf, K.P., Pironneau, O.: Pressure boundary conditions for blood flows. Chin. Ann. Math. Ser. B **36**(5), 829–842 (2015)
13. Hesthaven, J.S., Rozza, G., Stamm, B.: Certified reduced basis methods for parametrized partial differential equations. In: Springer Briefs in Mathematics, 1 edn. Springer, Switzerland (2015)
14. Janela, J., Moura, A., Sequeira, A.: Absorbing boundary conditions for a 3d non-Newtonian fluid-structure interaction model for blood flow in arteries. Int. J. Eng. Sci. **48**(11), 1332–1349 (2010)
15. Kunisch, K., Volkwein, S.: Galerkin proper orthogonal decomposition methods for parabolic problems. Numerische Mathematik **90**(1), 117–148 (2001)
16. Lombardi, M., Cremonesi, M., Giampieri, A., Parolini, N., Quarteroni, A.: A strongly coupled fluid-structure interaction model for wind-sail simulation. In: 4th High Performance Yacht Design Conference 2012, HPYD 2012 (2012)
17. Lombardi, M., Parolini, N., Quarteroni, A.: Radial basis functions for inter-grid interpolation and mesh motion in FSI problems. Comput. Methods Appl. Mechan. Eng. **256**, 117–131 (2013)
18. Maday, Y., Ronquist, E.M.: A reduced-basis element method. J. Sci. Comput. **17**(1), 447–459 (2002)
19. Malossi, A.C., Blanco, B.J., Deparis, S., Quarteroni, A.: Algorithms for the partitioned solution of weakly coupled fluid models for cardiovascular flows. Int. J. Numer. Meth. Biomed. Eng. **27**, 2035–2057 (2011)
20. Mameli, A., Magnusson, A., Aksenov, A., Kuznetsov, K., Luniewski, T., Moskalev, I.: Simulating the pouch forming process using a detailed fluid-structure interaction. In: Proceedings of the 2013 SIMULIA Customer Conference, pp. 1–15 (2013)
21. Olsson, M., Magnusson, A., Prasad, S.C.: Simulation of the forming process of liquid filled packages using coupled Eulerian-Lagrangian approach. In: Proceedings of the 2009 SIMULIA Customer Conference, pp. 1–12 (2013)
22. Quarteroni, A., Formaggia, L.: Mathematical modelling and numerical simulation of the cardiovascular system. In Ayache, N. (Guest ed.), Ciarlet, P.G. (ed.) Computational Models for the Human Body. Handbook of Numerical Analysis series, vol. XII. Elsevier (2004)
23. Quarteroni, A., Manzoni, A., Negri, F.: Reduced Basis Methods for Partial Differential Equations: An Introduction. Springer International Publishing, UNITEXT (2015)
24. Rathinam, M., Petzold, L.R.: A new look at proper orthogonal decomposition. J. Numer. Anal. **41**(5), 1893–1925 (2004)

Reduced-Order Modeling in the Manufacturing Process of Wire Rod: Applications for Fast Temperature Predictions and Optimal Selection of Process Parameters

Elena B. Martín, Fernando Varas and Iván Viéitez

Abstract

The number of operational variables that determines the cooling process of steel wire, given by the conveyor velocity and the different fan sections powers (controlled independently), lead to a dependency of the cooling on a high multidimensional parameter space whose potential combinations are impossible to be analyzed, either experimentally or by numerical simulation of a thermal–metallurgical model. To tackle this problem, an efficient strategy, based on the use of Higher Order Singular Value Decomposition (HOSVD), is presented. The approach presented provides a Reduced-Order Model (ROM) capable of predicting quite accurately the cooling curve for any combination of the process parameters. Fast online predictions of the cooling rates allow to incorporate accurate modeling results in many Engineering tools, such as model predictive control algorithms or plant simulation software. Also, the ROM in combination with an optimization tool

E. B. Martín (✉) · I. Viéitez
Consorcio Instituto Tecnológico de Matemática Industrial (ITMATI), Colegio de San Xerome
Praza Obradoiro, s/n, 15782 Santiago de Compostela, Spain
e-mail: emortega@uvigo.es

I. Viéitez
e-mail: ivvieitez@uvigo.es

E. B. Martín
Departamento de Ingeniería Mecánica, Máquinas y Motores Térmicos y Fluidos,
Universidad de Vigo. Campus Marcosende, 36310 Vigo, Spain

F. Varas
Departamento de Matemática Aplicada a la Ingeniería Aeroespacial, ETS de Ingeniería
Aeronáutica y del Espacio, Universidad Politécnica de Madrid, Plaza Cardenal Cisneros, 3,
28040 Madrid, Spain
e-mail: fernando.varas@upm.es

I. Viéitez
Departamento de Matemática Aplicada II, Universidad de Vigo, Campus Marcosende,
36310 Vigo, Spain

P. Quintela Estévez et al. (eds.), *Advances on Links Between Mathematics and Industry*, SxI - Springer for Innovation / SxI - Springer per l'Innovazione 15,
https://doi.org/10.1007/978-3-030-59223-3_4

finds the adequate operational parameters with significant reduction of energy consumption.

1 Introduction

The demand for higher quality materials in combination with more strict pollutants regulations and higher energy costs, force the steel industry to control efficiently the material processes, reducing, at the same time, consumption rates.

Industrial manufacturing of wire rod involves a hot rolling process, a subsequent coiling at high temperature and a final air-cooling generated by fans located beneath a conveyor. The final mechanical properties of the rod depend not only on the chemical composition of the steel but also on the cooling rates underwent by the product, which determines the final steel microstructure and, consequently, the final mechanical properties of the product [1].

Within this context, the use of predictive tools based on mathematical models that describe the dominant physics of the problem helps the industry to select the optimal combination of parameters for different types of processes and/or problems, assuring the quality of the outcomes. We present here a model-based tool for the fast prediction of the microstructure and temperature during the manufacturing industrial air-cooling process of wire rod. Starting from a coupled thermal–metallurgical model (solved by Finite Element Methods FEM) that has been previously validated for this type of process by comparison to experimental data [2,3], we describe the steps followed to construct the surface-response model (also named surrogate or ROM model) that allows a fast prediction of the wire cooling under different cooling conditions with a very good accuracy [3]. This model was obtained using HOSVD [4] in combination with Gappy–HOSVD techniques [5,6] to exploit efficiently a numerical database generated by the thermal–metallurgical model. Once the surrogate model is presented, its potential for process parameters design when used in combination with an optimization tool is shown.

The structure of this work is the following: Sect. 1.1 describes the characteristics of the analyzed industrial process, while Sect. 2 summarizes the numerical model (thermal and metallurgical model are in Sects. 2.1 and 2.2, respectively, while the implementation is described in 2.3). The ROM model is explained in Sect. 3. Finally, Sects. 4 and 5 present the applications of the developed tool for the fast prediction and the process parameters design. The conclusions are compiled in Sect. 6.

1.1 Description of the Industrial Process

The analyzed wire rod installation is located in Celsa Atlantic[1] company, at A Laracha (A Coruña), Spain. The coiled wire is formed in the laying head (see Fig. 1 left) at a temperature T_{LH} of approximately 1180 K (910 °C). The wire is moved along the

[1] http://www.celsaatlantic.com.

Fig. 1 Photograph of the laying head (left) and the conveyor belt (right). Reproduced with courtesy of Celsa Atlantic

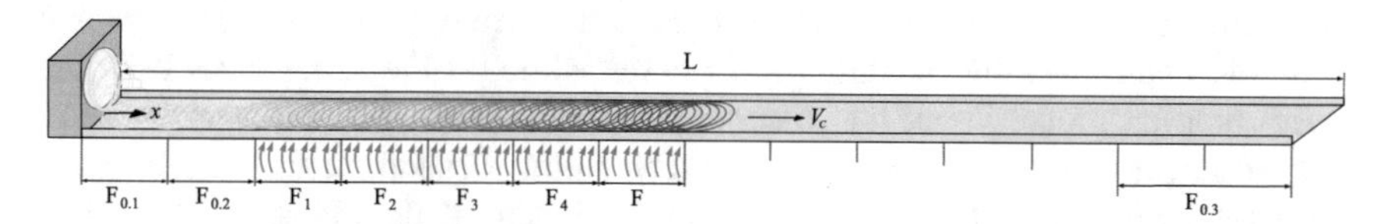

Fig. 2 Sketch of the installation and the different fans sections

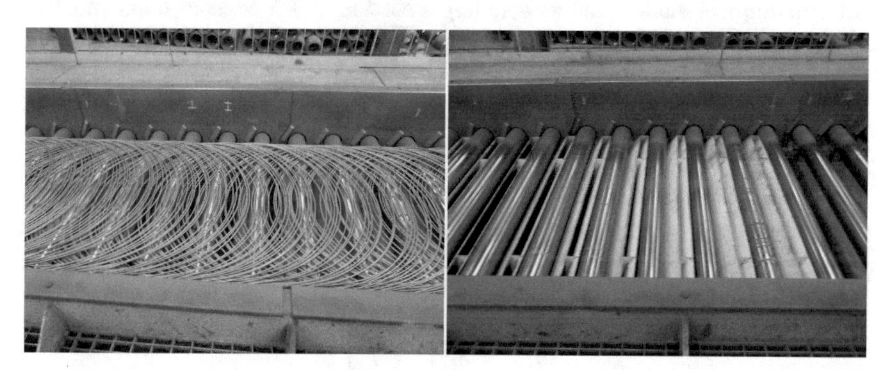

Fig. 3 Pictures of the stacking of the wire on the conveyor (left) and detail of the outlet fan sections (right). Reproduced with courtesy of Celsa Atlantic

conveyor (Fig. 1 right) by rotating cylindrical rods at a velocity V_c in the advancing x-direction. An sketch of the installation can be seen in Fig. 2. The conveyor, of length $L = 65$ m, has 14 different sections, marked in Fig. 2. Multiple centrifugal fans are located underneath some sections of the conveyor. The arrangement of the different groups of fans is indicated in Fig. 2. During production, fan groups F0.1 and F0.2 are generally switched off while the final group F0.3 is only used to cool the material sufficiently in order to be handled safely.

Each fan group F1 to F5 consists of three-speed adjustable centrifugal fans located under each section, one at the center and two at the edges. The outlet effective area of each fan can be seen in the right picture of Fig. 3. As in the lateral zone the coil wire is more densely stacked (see left picture 3), a higher air velocity should be given to that conveyor area. These centrifugal fans are regulated by adjusting the r.p.m. speed n to a variable torque given by the characteristic fan performance curve. Thus, estimation of the effective velocity of the air V_F given by each fan under the conveyor was done for the power ratio of each fan section $F_j = P_j/P_{nom}$. The estimated maximum velocity V_F for a 100% of power given by each of the lateral fans of groups $F1$ to $F5$ was 36 m/s for this specific installation.

The material analyzed in this work was T10 steel with a percentage of carbon $\%C$ of 0.08. Typically, three different product diameters D, equal to 5.5, 7 and 14 mm, are manufactured for this steel.

2 Numerical Model

The thermal and the microstructure evolution are strongly coupled. The thermal evolution determines the metallurgical phases while, at the same time, both the global thermal properties of the material (thermal conductivity and specific heat) and the latent heats released during transformations depend on the metallurgical phases, therefore, affecting strongly the thermal state of the material. Thus, they should be solved simultaneously. In the following Sects. 2.1 and 2.2 the thermal and metallurgical models and its coupling are detailed. All constants and variables involved in the previous equations are expressed, unless otherwise indicated, in the international system of units.

2.1 Thermal Model

Assuming negligible axial heat conduction, radial symmetry and uniform initial temperature, the piece temperature distribution $T(t, r)$ in the radial direction r follows:

$$\rho c_p \frac{\partial T}{\partial t} - \frac{1}{r}\frac{\partial}{\partial r}\left[rk\frac{\partial T}{\partial r}\right] = Q \tag{1}$$

where $r \in [0, D/2]$, being D the wire diameter. t is the time variable. The thermal conductivity k and specific heat c_p of the rod material depend on the proportion of each micro-constituent phase i, X_i, where i can be, depending on the piece thermal–metallurgical state, austenite, γ, (named hot phase) or ferrite, f, pearlite, p, bainite, b, and martensite, m, (also named cold phases). The different proportions X_i will be obtained after solving the metallurgical model, described below in Sect. 2.2.

$$k(T) = \sum_{i=\gamma,f,p,b,m} k_i(T)X_i \tag{2}$$

$$c_p(T) = \sum_{i=\gamma,f,p,b,m} c_{p,i}(T)X_i \tag{3}$$

For the steel material, density was made dependent on temperature in the following way:

$$\rho(T) = 7800 - 0.35(T - 273) \tag{4}$$

In Eq. (1), the heat source term Q represents the heat generated by metallurgical transformations during the cooling process and depends on the transformation rates $\partial X_i/\partial t$ and on the enthalpy of solid phase change ΔH_i:

$$Q = \rho(T) \sum_{i=f,p,b,m} \Delta H_i \frac{\partial X_i}{\partial t} \tag{5}$$

Equation (1) is solved with the initial uniform temperature $T(0,r) = T_{LH}$, the usual symmetry boundary condition at $r = 0$, and the following boundary condition at the rod surface $r = D/2$:

$$k\frac{\partial T}{\partial r} = C_a \left[\epsilon\sigma(T_\infty^4 - T^4) + \overline{h}_c(T_\infty - T)\right] + q_{LH} \tag{6}$$

as in [3]. The first right hand terms account for the heat losses by radiation and convection mechanisms respectively. Both effects are modulated by the wire stacking coefficient C_a across the conveyor, estimated as 0.8 for the lateral of the conveyor.

For the convective heat transfer coefficient $\overline{h}_c$, both free and forced convection were considered using available Nusselt number correlations for horizontal cylinders. In particular:

$$\overline{h}_c = \frac{k_{air}}{D}\left((\overline{Nu}_{nc})^3 + (\overline{Nu}_{fc})^3\right)^{1/3} \tag{7}$$

being k_{air} the air thermal conductivity at T_∞ of $298K$. The Nusselt number for natural convection $\overline{Nu}_{nc}$ is estimated as follows:

$$\overline{Nu}_{nc} = C Ra^p, \quad Ra = Pr\frac{gD^3}{\nu_{air}^2}\frac{T - T_\infty}{T} \tag{8}$$

Constant parameters C and p for the natural convection depend on the Rayleigh value Ra, as indicated in [3].

For the forced convection, $\overline{Nu}_{fc}$ is obtained from:

$$\overline{Nu}_{fc} = C Re^p Pr^{1/3}; \quad Re = \frac{V(t)D}{\nu_{air}} \tag{9}$$

where the selected constant values of C and p depend on the range of the Reynolds number Re [3]. Air Prandtl number Pr in Eqs. (8) and (9) was taken as 0.7. The velocity of the surrounding air $V(t)$ was calculated as a composition of the conveyor velocity V_c and the velocity induced by the fans beneath it V_F, that is,

$$V(t) = \sqrt{V_c^2 + V_F^2(x)} \tag{10}$$

where x is the distance-coordinate to the laying head along the conveyor (indicated in 2), which coincides with the wire rod advancing direction. Thus $x = V_c t$. As explained in Sect. 1.1, $V_F(x)$ depends on the corresponding fan group operational power ratio F_j for each section of the conveyor.

The last term q_{LH} of Eq. (6) represents the heat transmitted by radiation of the laying head to the neighboring wire, whose effects cannot be neglected, especially at the beginning of the run-out table. This heat transfer was modeled in [3] by the following equation:

$$q_{LH} = \epsilon \sigma T_{LH}^4 e^{-x/L_c} \tag{11}$$

The value of the length parameter L_c was taken as 5 m and it was adjusted with experimental measurements in previous works [2]. Heat conduction losses associated to the contact between the wire and the conveyor rollers were not taken into account in the model.

2.2 Metallurgical Model

The transformations dominated by carbon diffusion processes (that potentially lead to ferrite f, pearlite p and bainite b phases from the initial austenite γ phase) have been modeled by an Avrami type equation, as described in [3]:

$$X_i = X_\gamma X_{i,max} \left[1 - exp\left(-b\left(\frac{d_\gamma}{d_{\gamma,ref}}\right)^m t^n\right)\right] \tag{12}$$

where X_i stands for the proportion of micro-constituent i, $X_{i,max}$ is the maximum proportion of micro-constituent i at a given temperature, and X_γ is the proportion of austenite at the beginning of each transformation. Material parameters b and n are extracted from the isothermal transformation (TTT) diagram for the carbon steel AISI 1008. Grain size d_γ was taken as 8 ASTM (22.1 μm) (information provided by Celsa Atlantic). For continuous cooling, the additive rule of Scheil that discretizes the cooling in small intervals of constant temperatures has been applied. The beginning of the transformations is determined by

$$\sum \frac{\Delta t_i}{\tau_i\left(\frac{d_\gamma}{d_{\gamma,ref}}\right)} \geq 1 \tag{13}$$

where τ_i stands for the incubation time for a specific microstructure for each temperature, given by the TTT diagram.

The martensitic transformation is determined by the temperature reached during the heat treatment. The Koïstinen–Marburger model establishes the following algebraic equation (14) for the martensite proportion X_m

$$X_m = \left(1 - X_f - X_p - X_b\right)\left(1 - e^{\beta[M_s - T]^+}\right) \tag{14}$$

being β a material parameter, M_s the temperature for the beginning of the martensitic transformation (extracted from the continuous cooling transformation (CCT) diagram of the Steel AISI 1008).

2.3 Implementation and Validation of the Model

The numerical model described in Sects. 2.1 and 2.2 was discretized using a Finite Element Method (FEM) programmed in Python[2]. P1 polynomials were used for the spatial discretization of the temperature field, while an Euler implicit scheme was selected for temporal discretization. The resulting problem was solved on a uniform mesh of $\Delta r = 0.1$ mm with a constant time step of $\Delta t = 0.1$ s. For coherence between simulations at different conveyor velocities V_c, additional time steps were added to obtain the wire rod temperature at the same specific positions x for all the computations: the beginning, the middle and the end of each of the 14 sections of the conveyor.

Validation of the model, by comparison between the wire surface temperatures extracted from the numerical results and experimental measures for three different wire rod diameters D (5.5, 7 and 14 mm) under different cooling conditions, was carried out in [3]. Averaged relative errors in temperature for the complete test cases were below 5%, enabling the numerical tool for the prediction of the temperature and metallurgical evolution of the wire during the process.

Therefore, for a given combination of the process parameters (inputs), that is, power ratio of each fan group F_i and the conveyor velocity V_c, this model provides the temperature evolution of the wire $T(x, r)$ (output).

As an example, for the material diameter of 14 mm, the discretized (output) $T(x, r)$ comprises a total of 2059 values (29 x-positions by 71 r-nodes) for each combination of the inputs process parameters. Computational time of one (non-parallelized) case for the complete model (in an Intel Core i7 2.67 GHz–12 GB RAM computer) took approximately 60 s.

[2]www.python.org.

Coupled thermal–metallurgical model:
Inputs: F_1, F_2, F_3, F_4, F_5 and V_c.
Outputs: $T(x, r)$, $29 \times N$ values (29 x-positions by N r-nodes from FEM numerical solution)

3 Reduced-Order Model

The final goal of the developed model is to provide very fast predictions of the wire cooling to be used in different contexts. For instance, these predictions would be used during the design of the operation conditions for a new product. In this case, either the cooling line or the whole manufacturing line is to be simulated for each set of operation parameters and the computational cost of each numerical simulation must be very low if a large number of configurations are to be explored. Also, if a model predictive control (MPC) strategy is to be implemented in the cooling line, a fast online cooling prediction will be needed in order to provide model predictions to the controller.

Certainly, computational speed of the numerical model presented in the previous section can be greatly improved in many ways. First, the use a compiled version of the code (rewritten in Fortran or C++) would speed up significantly the numerical prediction (with the use of wrappers, this new code could still be called from Python if this language is intended to be used in design tasks, as it is beginning to be customary in many industrial fields). Additional speed up could come from the use of parallelization. In this sense, cooling prediction for several sets of line operation conditions represents an embarrassingly parallel problem, of course. Numerical simulation of a single case, instead, offers few opportunities to (effective) parallelization due to the small scale of the discrete problem.

Nevertheless, our approach will be quite different. Instead of accelerating the numerical simulations through the code improvement, a reduced-order model will be used. As this type of approach can be applied (in principle) to any numerical model, it is very useful (even essential) when dealing with very time-consuming numerical simulations (e.g., 3D numerical FEM and/or CFD simulations), which makes its use very appealing in different industrial problems. Following this approach, in an offline (or preprocessing) phase a number of numerical predictions are computed. Then, the information from these predictions is processed in order to derive a fast, reduced-order model able to provide almost instantaneous cooling predictions in an online phase (either in a design environment or in a model predictive controller, according to the previous examples).

In particular, our reduced-order model will proceed along these lines in the offline phase:

1. Build a database of cooling predictions for several combinations of the six cooling line process parameters: conveyor velocity and power ratio of five fan groups.
2. Store this database in a seven-dimensional tensor: six entries in the tensor will correspond to the process parameters and the seventh one to the solution components (typically, surface temperature for some prescribed positions along the cooling curve, but if a more detailed description of the material cooling is needed all the values of the numerical solution can be stored).
3. Complete all the entries in the seven-dimensional tensor (since computed cooling predictions will not cover the whole set of entries in this high-dimensional tensor).
4. Prepare database information in order to easily interpolate a cooling curve for any combination of the six cooling line process parameters.

As a result of this offline phase, a tool for fast cooling prediction is developed. Ideally, to compute a cooling curve with this tool would take at most a few milliseconds.

Concerning the development of this reduced-order model, a high order singular value decomposition (or HOSVD, see [4]) will be used to decompose the tensor representing the complete database A in terms of a core tensor S (to be later compressed) and orthogonal matrices U^i associated to each parameter in the database.

$$a_{i_1 i_2 \ldots i_N} = \sum_{j_1=1}^{I_1} \sum_{j_2=1}^{I_2} \cdots \sum_{j_N=1}^{I_N} s_{j_1 j_2 \ldots j_N} u^{(1)}_{i_1 j_1} u^{(2)}_{i_1 j_1} \ldots u^{(N)}_{i_1 j_1} \tag{15}$$

HOSVD decomposition provides a tool to perform data compression since the core tensor S can be truncated, leading to a smaller database. At the same time (and, in a sense, more importantly) orthogonal matrices U^i can be used to propose an easy procedure to interpolate in any parameter (through interpolation of the matrices elements). Details of this procedure can be found in [3].

Additionally, a technique is needed to build a complete tensor from a small set of tensor elements (otherwise a process with a relatively large number of operation parameters would lead to an unaffordable computational workload to compute all the database entries). A technique inspired by the so-called Gappy–POD proposed in [5] and able to deal with HOSVD decompositions can be found in [6].

The last ingredient in the derivation of a surrogate model is the proposal of a sampling technique to select a small fraction of tensor entries retaining enough information for the Gappy–POD to be able to complete the whole database with a high accuracy. Such a technique is presented in [3].

4 Fast Prediction of Cooling Curves

Following the methodology outlined in the previous section (and detailed in [3]), we present here an application of this tool.

In this example, we will consider the preparation of a database with five values of the power ratio (0, 25, 50, 75 and 100%) for each group of fans and three values of the conveyor velocity (20, 30 and 40 m/min). Using this database, a fast tool to obtain cooling curve predictions will be developed.

The computation of the whole database is clearly to be avoided. This database would contain 9375 numerical predictions of the cooling curve. Even with an optimized version of the code, this computation would take many hours. Using our Python code, computation time would tot up 5 days if it is sequentially executed although, as mentioned, it is an embarrassingly parallel task and can be carried out in a shorter time using a computer cluster.

Instead, the use of a Gappy–POD technique would allow the generation of the complete database from a reduced number of entries, thus completely avoiding the need of advanced computer resources or a very large computing time in the offline phase. In particular, in this example, a (much) smaller database of 1278 cases is computed (around 13% of the total cases in the full database) using a case selection strategy described in [3].

From this small fraction of the full database, the 9375 numerical predictions for all the combinations of the operation parameters can be obtained using the Gappy–HOSVD technique described in [6], the database compressed and information from the HOSVD decomposition preprocessed to allow fast interpolation for any arbitrary combination of operation parameters.

The offline computational tool implementing all these ideas is able to make a new prediction in less than 5 ms then fulfilling the requirements to be implemented in design environments or MPC controllers (among many other Engineering applications).

Gappy-HOSVD:

- Initialization: generation of the gappy database (1278 cases): 22 h
- Reconstruction of the complete database (9375 cases): 75 min
- Prediction for any inputs combination (F_1, F_2, F_3, F_4, F_5, V_c): < 5 ms
- Temperature errors: Maximum = 12 K, averaged = 0.75 K

Two examples of fast prediction are presented here. Both correspond to a rod wire with a 14 mm diameter made of a T10 steel. The first one (see top plot of Fig. 4) compares reconstructed temperature profile for a case not included in the database discretization ($F_1 = 68\%$, $F_2 = 40\%$, $F_3 = 30\%$, $F_4 = 20\%$, $F_5 = 20\%$, $V_c = 25\,\mathrm{m/min}$) to the thermal prediction made by the complete FEM model. The maximum error is less than 5 °C. For the second example (bottom plot of Fig. 4) some groups of fans are switched off ($F_1 = 0\%$, $F_2 = 20\%$, $F_3 = 0\%$, $F_4 = 80\%$, $F_5 = 20\%$, $V_c = 40\,\mathrm{m/min}$). Also, the errors in temperature are very small (less than a 6 °C for the maximum error).

To show the errors of the developed tool, the reconstruction of the complete database was carried out for all the studied product diameters. Table 1 shows the maximum temperature deviation (from the FEM computed cooling curve) for the

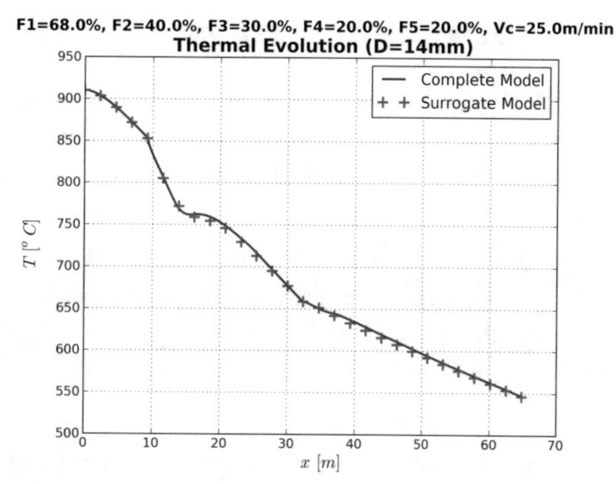

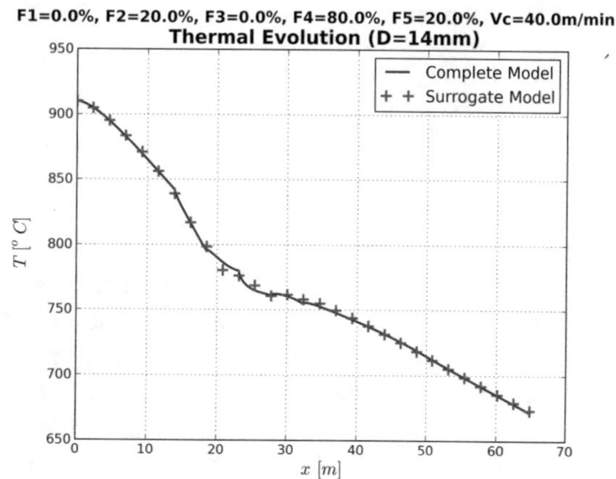

Fig. 4 Surface temperature evolution ($D = 14$ mm) predicted by the FEM numerical model and by the reduced-order model for two different cases with six operational parameters

Table 1 Maximum temperature errors of the worst reconstructed cases

Diameter (mm)	Max. error (K)	Operational parameters of the worst reconstructed cases
5.5	19.5	$F_1 = 100\%$, $F_2 = 0\%$, $F_3 = 100\%$, $F_4 = 25\%$, $F_5 = 75\%$, $V_c = 40$ m/min
7	8.5	$F_1 = 50\%$, $F_2 = 0\%$, $F_3 = 0\%$, $F_4 = 0\%$, $F_5 = 75\%$, $V_c = 40$ m/min
14	11.7	$F_1 = 0\%$, $F_2 = 0\%$, $F_3 = 0\%$, $F_4 = 100\%$, $F_5 = 25\%$, $V_c = 40$ m/min

worst reconstructed cases in the complete database. These values provide maximum relative errors in temperature below 5%. As can be seen, the worst reconstructed cases are located in the boundaries of the parameter space.

5 Optimization of Process Parameters

The developed surrogate model can be used combined with an optimization algorithm. The purpose is to find out, with a very reduced computational time, the best operational parameters of the process according to a specific criterion:

- To minimize the deviation from a reference cooling curve.
- To reduce the energy consumption.

To fulfill simultaneously both demands, the following functional, shown in equation (16), was built

$$\min_{\mathbf{x}} \quad \omega_1 \hat{\theta}_1 \sum_{n=1}^{N} \sum_{i=1}^{29} \left(|T^{ref}_{i,n} - T_{i,n}(\mathbf{x})| \right)^2 + \omega_2 \hat{\theta}_2 \sum_{j=1}^{5} F_j P_{nom} L / V_c \tag{16}$$

The first term of the functional (16) measures the deviation from the reference temperature curve, given by T^{ref}, at specific locations in the conveyor and mesh $r-$nodes. The second-term measures the energy consumption, that is, the power of each fan P_j (given by the power ratio F_j and the nominal power P_{nom}) and the residence time of the wire on the conveyor L/V_c. Parameters ω_1 and ω_2 are the weights of each term of the functional, while $\hat{\theta}_1$ and $\hat{\theta}_2$ are used to make the contribution of each term dimensionless. $\mathbf{x}$ is the input parameters vector, constituted by the different power ratios and the conveyor velocity:

$$\mathbf{x} = (F_1, F_2, F_3, F_4, F_5, V_c) \tag{17}$$

The minimization problem, that is, minimization of the functional $f(\mathbf{x})$ given by Eq. (16), was solved using the ROM combined with a Trust Region Method. The main ideas of Trust Region Method will be stated below (details of the Trust Region Method can be consulted in [7–9], while an application to the steel industry can be seen in [10]).

From a given a value $\mathbf{x}_k$, the iterative loop to find the next iterate $\mathbf{x}_{k+1}$ follows these steps:

1. Assembling of a quadratic model of the objective functional (generally the Taylor polynomial) for $\mathbf{x}_{k+1} - \mathbf{x}_k = \mathbf{p}$:

$$m_k(\mathbf{p}) = f_k + \mathbf{g}_k^T \mathbf{p} + \frac{1}{2} \mathbf{p}^T B_k \mathbf{p} \tag{18}$$

2. Obtaining of the minimum $\mathbf{p}_k$ of the previous quadratic model over the trust region Δ_k ($\|\mathbf{p}_k\| < \Delta_k$).
3. Updating the size of the trust region depending on the quality of the approximation of f provided by m_k.
4. Accepting or rejecting the new calculated iterate $\mathbf{x}_k + \mathbf{p}$ depending on the functional reduction. If it is rejected, the step is solved again with the new trust region size.
5. Stopping criterion: $\|\nabla f(\mathbf{x}_{k+1}\| < \tilde{\epsilon}$.

As an example of the application of the optimization method, a reference temperature curve was created from the FEM computations of the physics-based model of wire rod cooling (for $D = 14$ mm) with the following process parameters: $F_1 = 79\%$, $F_2 = 79\%$, $F_3 = 28\%$, $F_4 = 28\%$, $F_5 = 28\%$ and $V_c = 34$ m/min. This is equivalent to regroup the first to groups F_1 and F_2 in one, $G_1 = 79\%$, and the rest, F_3, F_4 and F_5, in a second one $G2 = 28\%$. Assuming the following weights $\omega_1 = 0.7$ and $\omega_2 = 0.3$, the optimization for the input parameters was obtained in less than 2 s.

Example of optimization of input parameters:

- T^{ref} given by FEM for input process parameters $G_1 = 79\%$, $G2 = 28\%$, $V_c = 34$ m/min.
- Input parameters given by minimization of (16): $G_1 = 41.6\%$, $G2 = 12.3\%$, $V_c = 31$ m/min.

The results are shown in Fig. 5, where the reference curve T^{ref} for the wire surface obtained from the FEM simulation of the complete model is plotted in blue color while the cooling temperature results provided by the optimized input parameters ($G_1 = 41.6\%$, $G2 = 12.3\%$, $V_c = 31$ m/min) is plotted with red crosses. The maximum error between them is less than 5 °C, therefore, both combinations of inputs parameters produce, approximately, the same cooling curve. However, the consumption reduction between those two different selections of input parameters is of a 45%.

Another example is shown in Fig. 6, where a reference temperature curve (for $D = 14$ mm) was created from the FEM computation with the following process parameters: $F_1 = 77\%$, $F_2 = 45\%$, $F_3 = 35\%$, $F_4 = 15\%$, $F_5 = 20\%$ and $V_c = 34$ m/min. Assuming again the previous weights ω_1 and ω_2 in the minimization function, the optimization for the input parameters was also obtained in less than 2 s, providing a maximum error in temperature of less than 15 °C. The selected process parameters given by the optimization were $F_1 = 61.8\%$, $F_2 = 33.8\%$, $F_3 = 23.8\%$, $F_4 = 13\%$, $F_5 = 13\%$ and $V_c = 33.9$ m/min, which produce an energy consumption reduction of a 25%.

6 Conclusions

This work presents two different applications of an efficient computational tool, based on Gappy–HOSVD techniques, to predict the temperature evolution of wire rod cooling. The studied industrial process has up to 6 operational parameters: each power of the cooling fan sections and the conveyor velocity that carries the coil rod.

First, a previously validated FEM thermo–metallurgical model giving the wire temperature evolution was presented. Each of the predictions of the numerical implementation of this physics-based model took approximately 60 s. Then, a reduced number of combinations of the 6 operational parameters cases (a 13% of the total of potential combinations) were computed to construct a reduced database that was exploited with Gappy–HOSVD techniques. The computational time taken for the implementation of these techniques was 75 min (offline phase) while the prediction

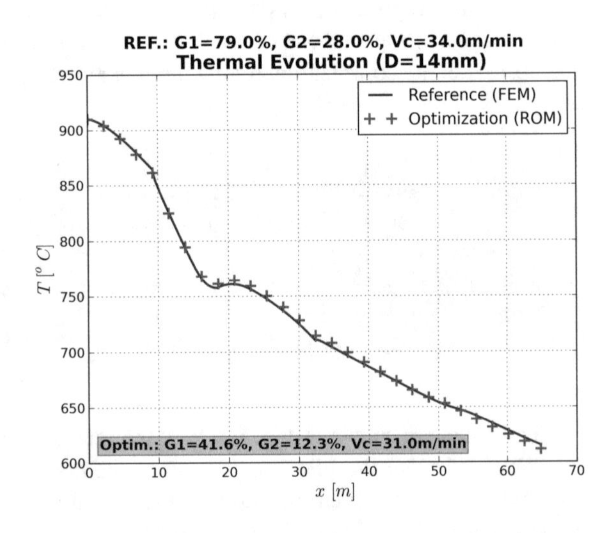

Fig. 5 Example 1: Surface temperature evolution ($D = 14$ mm) obtained by the surrogated model combined with the Trust Region Method, compared to the cooling reference curve

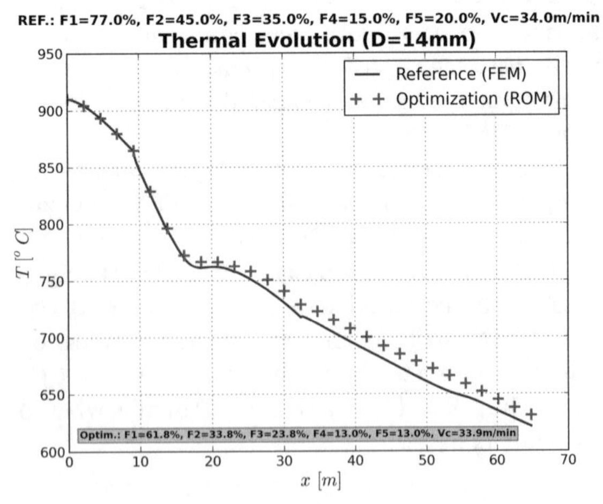

Fig. 6 Example 2: Surface temperature evolution ($D = 14$ mm) obtained by the surrogated model combined with the Trust Region Method, compared to the cooling reference curve

of any combination of operational parameters, using the information provided by the Gappy–HOSVD, took less than 5 ms (online phase). The performance in terms of temperature prediction provided a maximum temperature deviation for the worst predicted cases of less than 20 °C, 9 °C and 12 °C for the product diameters of 5.5, 7 and 14 mm, respectively, with averaged errors of less than 0.75 °C. Finally, results of the exploitation of this surrogate model in combination with a trust region optimization method were presented. The optimization of the two examples, which used the minimization of the deviation from a reference cooling temperature and the process energy consumption, was capable of selecting the appropriate process parameters in less than 2 s. Maximum temperature deviations from the reference cooling temperature were of 5 °C and 15 °C, respectively. Reduction in the energy consumption (when compared to the process parameters that produced the reference cooling temperature) was very significant, of a 45 % and a 25 % for the two examples presented.

The affordable computational time taken to build the surrogate model, and the very short time taken for the temperature prediction and the optimization of the process parameters, make the proposed technique suitable to be implemented in different tasks, that is, plant simulation software and model predictive control (MPC) algorithms.

Next steps in our research deal with the extension of this methodology to industrial processes with a larger number of control variables. In particular, we are currently modeling a paper resin impregnation and curing line in a lamination paper manufacturing plant, where typically between 10 and 20 process variables can be modified in order to control the product quality (considering several raw materials and desired product properties, as well as operating under different ambient variables).

As shown in this paper, the presence of a large number of input variables in a reduced-order model (which invariably results in difficulties in this framework, quite expressively described by the "curse of dimensionality") makes difficult the implementation of the methodology described above. Thus, it seems advisable to turn our attention to more flexible approaches. In particular we are considering the use of "moving least squares" techniques for scattered data interpolation [11] to deal with such problems.

Acknowledgements This research was supported by the Ministry of Economy and Competitiveness (Spain) (grant TRA2016–75075-R) and the Centre for the Development of Industrial Technology (CDTI), Spain, in collaboration with companies Celsa Atlantic S.L. and Russula S.A.U. (grant ITC-20113017).

References

1. Anelli, E.: Application of mathematical modelling to hot rolling and controlled cooling of wire rods and bars. ISIJ International **32**(3), 440–449 (1992)
2. Viéitez, I., López-Cancelos, R., Martín, E., Varas, F.: Predictive model of wire rod cooling. In: Proceedings of the 28th ASM Heat Treating Society Conference, pp. 518–524 (2015)

3. Viéitez, I., Varas, F., Martin, E.: An efficient computational technique for the prediction of wire rod temperatures under different industrial process conditions. App. Therm. Eng. **149**, 287–297 (2019)
4. De Lathauwer, L., De Moor, B., Vandewalle, J.: A multilinear singular value decomposition. SIAM J. Matrix Anal. Appl. **21**(4), 1253–1278 (2000)
5. Everson, R., Sirovich, L.: Karhunen-Loeve procedure for gappy data. J. Opt. Soc. Am. A **12**(8), 1657–1664 (1995)
6. Moreno, A., Jarzabek, A., Perales, J., Vega, J.: Aerodynamic database reconstruction via gappy high order singular value decomposition. Aerosp. Sci. Technol. **52**, 115–128 (2016)
7. Nocedal, J., Wright, S.J.: Numerical Optimization, 2nd edn. Springer Science Business Media (2006)
8. Conn, A.R., Gould, N.I.M., Toint, P.L.: Global convergence of a class of trust region algorithms for optimization with simple bounds. SIAM J. Numer. Analy. **25**(2), 433–460 (1988)
9. Steihaug, T.: The conjugate gradient and trust regions in large scale optimization. SIAM J. Numer. Analy. **20**(3), 626–637 (1983)
10. Martín, E., Meis, M., Mourenza, C., Rivas, D., Varas, F.: Fast solution of direct and inverse design problems concerning furnace operation conditions in steel industry. Appl. Therm. Eng. **47**, 41–53 (2012)
11. Fasshauer, G., Zhang, J.: Scattered data approximation of noisy data via iterated moving least squares. In: T. Lyche, J. L. Merrien and L. L. Schumaker (eds.) Proceedings of Curve and Surface Fitting: Avignon 2006. Nashboro Press, Brentwood, TN (2007)

Modeling and Numerical Simulation of the Quenching Heat Treatment. Application to the Industrial Quenching of Automotive Spindles

Carlos Coroas and Elena B. Martín

Abstract

The quenching heat treatment consists in the immersion of a steel piece (previously heated up to the austenization temperature range) in fluid. The fast cooling undergone by the piece induces microstructure transformations (from austenite to usually martensitic microstructure) aimed to provide the piece with specific mechanical properties (high hardness). The numerical model needed to mimic the cooling process and, therefore, to predict the final crystallographic structure, involves the following strongly coupled problems: a two-phase turbulent thermo-fluid-dynamic model (due to the presence of liquid and vapor caused by the high solid temperatures), and a thermal-metallurgical model for the piece. In this work, the heat flow on the surface of the spindle is characterized using a compilation of correlations (based on the specialized literature and also adjusted from simplified experiments and/or simulations) aimed to describe the different heat transfer mechanisms, extensively described in Nukiyama's experiments. This approach allows to describe (up to a certain degree of accuracy) the cooling process without solving a complex fluid-dynamic multiphase model, and hence in a computationally affordable way. The final model is eventually used to optimize the manufacturing parameters of the steel industrial quenching process of spindles in the automotive industry.

C. Coroas
Departamento de Matemática Aplicada II, E. de Ing. de Telecomunicación, Universidade de Vigo, Campus Marcosende, 36310 Vigo, Spain
e-mail: ccoroas@uvigo.es

E. B. Martín (✉)
Departamento de Ingeniería Mecánica, Máquinas y Motores Térmicos y Fluidos, E. de Ing. Industrial, Universidad de Vigo, Campus Marcosende, 36310 Vigo, Spain
e-mail: emortega@uvigo.es

Instituto Tecnológico de Matemática Industrial ITMATI, Colegio de San Xerome Praza Obradoiro, s/n, 15782 Santiago de Compostela, Spain

P. Quintela Estévez et al. (eds.), *Advances on Links Between Mathematics and Industry*, SxI - Springer for Innovation / SxI - Springer per l'Innovazione 15,
https://doi.org/10.1007/978-3-030-59223-3_5

1 Introduction

Industrial quenching is a heat treatment used on steel pieces aimed to provide them with desirable mechanical characteristics, particularly a high superficial hardness. This is achieved by obtaining a suitable microstructure during the heat treatment process. To induce these metallurgical transformations, a fast cooling is required. First, the steel pieces are heated up to attain the austenitic microstructure, and then, they are immersed in a tank of liquid, usually subjected to agitation. This process involves (i) the heat transfer between the solid piece and a dynamical two-phase fluid and (ii) the metallurgical transformation problem, which is strongly coupled to the thermal evolution of the piece.

Although the phase transformation problem can be described with well studied models, the heat transfer problem involves the resolution of different physical phenomena (due to the vapor generation during the immersion) with very different space and temporal scales [1–4]. This is a very complex problem that has not yet been tackled for the industrial quenching processes. To simplify the heat transfer between the piece surface and the two-phase fluid, literature correlations describing the different regimes encountered experimentally [5–7], or experimentally adjusted heat transfer coefficients are commonly used.

In this work, we present a thermal-metallurgical model capable of predicting accurately the final steel microstructure under different process parameters at a reduced computational cost which makes it suitable for the industrial process analysis. A heat transfer coefficient, which takes into account the different control parameters of the process, is modeled based on literature correlations and used to describe the heat transfer during all the regimes that take place during the cooling.

The work is organized as follows: the industrial quenching process is explained in Sect. 1.1. Then, Sect. 2 presents the numerical model, divided in Sects. 2.1 and 2.3 that explain, respectively, the thermal and the metallurgical model. Section 2.2 is devoted to the validation of the proposed thermal model with laboratory experiments carried out by the research group. Description of the numerical implementation of the complete model is summarized in Sect. 2.4. Finally, we analyze the results of the coupled thermal-metallurgical model in Sect. 3 while the main conclusions are shown in Sect. 4.

1.1 Industrial Quenching Process

An industrial quenching process carried out by the company CIE Galfor[1] (located in Ourense, Spain) is going to be analyzed. They produce truck axle spindles of length L and averaged thickness dimensions e of 285 mm and 20.75 mm, respectively. The spindles will act as joint devices between the wheels, the axle, and the suspension system of trucks. A high hardness is needed on the piece surface to support the

[1] https://www.cieautomotive.com/-/cie-galfor.

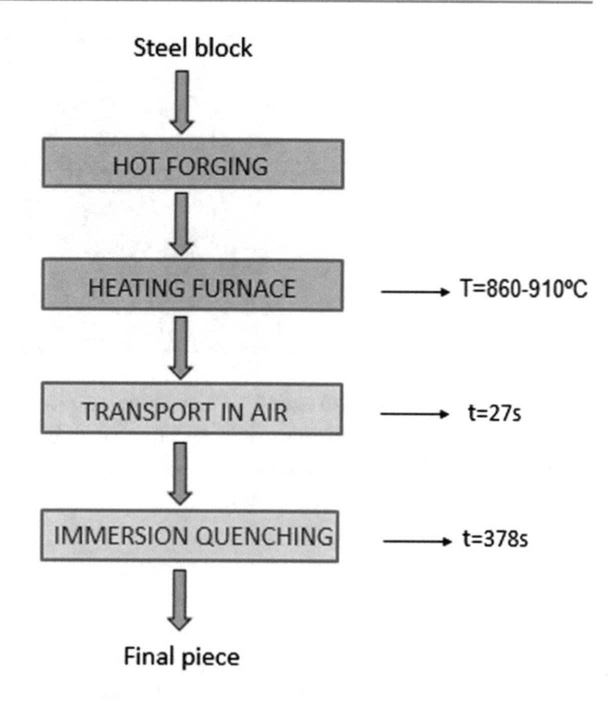

Fig. 1 Industrial process diagram which shows the different manufacturing stages

friction stresses during its use. The pieces material is an alloyed steel classified into the F-130 group with a 0.23% of carbon.

The manufacturing process consists of four stages as indicated in Fig. 1: a hot extrusion forging, a homogenization heating in a continuous furnace, a transport of the piece (surrounded by air) to the immersion tank and, finally, the immersion in the fluid bath. However, during this study, as we are interested in the metallurgical transformations during the cooling stages, only the last two stages are going to be analyzed. In consequence, a uniform homogenization temperature will be assumed at the exit of the pieces from the furnace. After the furnace, an automated mechanism elevates the pieces in batches of four and it transports them until the submersion on the non-pressurized water tank shown in Fig. 2. The transport time takes around 27 s. The piece is kept inside the fluid during approximately 378 s. The complete immersion of the pieces in the fluid takes around 4 s.

The goal of this study is the design of a numerical tool capable of predicting the final microstructure acquired during this cooling process. This allows to evaluate the influence that the different process parameters, namely,

- the steel chemical composition,
- the homogenization temperature of the furnace,
- the waiting time surrounded by air,
- the quenching agitation velocity, and
- the fluid temperature

Fig. 2 Liquid tank used for the quenching of truck spindles, provided by company CIE Galfor

have on the final microstructure of the spindle. This tool prevents the realization of lots of costly experimental trials and improves the efficiency of the process by analyzing the adequacy (in terms of the final microstructure of the pieces) of different combinations of process parameters.

2 Numerical Model

As indicated previously, the thermal problem is strongly coupled to the metallurgical changes. The thermal evolution determines the microstructure, while the thermal properties, at the same time, depend on the instantaneous proportion of each metallurgical phase. Thus, these models have to be solved with a two-way coupling approach.

During the description of the equations below, all constants and variables are expressed in the international system of units.

2.1 Thermal Model

The equation to be solved for the temperature field T, in the piece domain, Ω, corresponds to the transient parabolic partial equation (1), with the heat transfer boundary condition at the fluid-piece contact surface, $\partial\Omega$, defined by Eq. (2):

$$\rho C_p \frac{\partial T}{\partial t} - div(k\nabla T) = Q \qquad in\ \Omega \times (0, t_f) \tag{1}$$

$$-k\nabla T \cdot \eta = q = h(T - T_{ref}) \qquad on\ \partial\Omega \times (0, t_f) \tag{2}$$

where q represents the heat flux (per unit of surface) that comes out from the piece to the surrounding fluid while t_f is the final time considered for the thermal model.

ρ is the temperature dependent material density,

$$\rho(T) = 7800 \left[\frac{kg}{m^3}\right] - 0.35 \left[\frac{kg}{m^3 K}\right] T, \tag{3}$$

while k and C_p are the thermal conductivity and specific heat, which vary with the temperature [8] and depend on the proportion X_i of the different phases i. The phases i can be either austenite, γ, called hot phase, ferrite, f, pearlite, p, bainite, b, and/or martensite, m, also called cold phases.

$$k(T) = \sum_{i=\gamma,f,p,b,m} k_i(T)X_i \tag{4}$$

$$C_p(T) = \sum_{i=\gamma,f,p,b,m} C_{p,i}(T)X_i \tag{5}$$

The heat source term Q represents the heat generated by the metallurgical transformations during the cooling process and depends on the transformation rates $\partial X_i/\partial t$ and on the enthalpy of solid phase change ΔH_i, which can be extracted from [9] and [10].

$$Q = \rho(T) \sum_i \Delta H_i \frac{\partial X_i}{\partial t} \tag{6}$$

The parameter h of Eq. (2) stands for the effective heat transfer coefficient, which includes all the heat transfer phenomena between the fluid and the piece surface, and it will be defined in the following sections for the two different stages of the cooling process (transport in air and water submersion).

2.1.1 Heat Transfer to Air

This stage is characterized by free convection to air (at a constant reference temperature, taken as $T_{ref} = 298$ K) and radiative heat losses.

The free convective heat transfer coefficient $h_c = \frac{k_{air}\overline{Nu}_{cyl}}{L}$ (where k_{air} stands for the thermal conductivity of the air and L is the characteristic piece length) is obtained from the following correlation for the Nusselt number (extracted from [11] for free convection in vertical cylinders):

$$\overline{Nu}_{cyl} = \left(0.825 + \frac{0.387 Ra_L^{\frac{1}{6}}}{[1 + (\frac{0.492}{Pr})^{\frac{9}{16}}]^{\frac{8}{27}}}\right)^2 (1 + 1.3\zeta^{0.9}) \tag{7}$$

In Eq. (7) Ra_L and Pr stand for the air Rayleigh and air Prandtl numbers, respectively. ζ is a curvature correction factor $\zeta = L/D\ Ra_L^{-1/4}$, being D the averaged spindle diameter.

In addition to the convective heat flux, the radiative heat flux, q_{rad}, and, in consequence, its correspondent heat transfer coefficient, will be modeled by a common Stefan-Boltzmann expression (8).

$$q_{rad} = \epsilon\sigma_{SB}\left({T_{ref}}^4 - T^4\right) \tag{8}$$

where σ_{SB} is the well-known Stefan-Boltzmann constant and ϵ is the surface temperature dependent steel emissivity [10]:

$$\epsilon = \frac{T}{1000}\left(0.125\frac{T}{1000} - 0.38\right) + 1.1 \tag{9}$$

2.1.2 Heat Transfer to the Quenching Fluid

Nukiyama's experiments [2] revealed that during pool boiling, the evolution of the heat flux between the piece surface and the liquid is complex due to the different evaporation regimes that the fluid will undergo. In this study, different functions are proposed to characterize the heat flux q during each regime, all of them dependent on the surface wall temperature T. These expressions are presented below, following the cooling transient sequence suffered by the piece during the quenching process.

Film boiling

During this regime, the piece is still at a very high temperature, so the wall is surrounded by a vapor blanket, which acts as an insulator between the piece surface and the liquid. In order to characterize this effect, a global heat transfer coefficient h will be defined considering the radiative heat transfer [6,12]:

$$h = h_c(1 + 0.025(T_{sat} - T_b)) + 0.75h_{rad} \tag{10}$$

being T_{sat} the fluid saturation temperature and T_b the constant bulk temperature of the quenching bath (measured far enough from the piece). Heat transfer coefficient h_{rad} stands for the radiation mechanism, following [6]:

$$h_{rad} = \frac{\sigma_{SB}(T^4 - T_{sat}^4)}{\left(\frac{1}{\epsilon_w} + \frac{1}{\epsilon_l} - 1\right)(T - T_{sat})} \tag{11}$$

where ϵ_l y ϵ_w are the fluid and the piece emissivities.

The convection coefficient h_c is defined following [12] as

$$h_c = 0.94\left[\frac{k_v^3\rho_v(\rho_l - \rho_v)h'_{fg}g}{L_c\mu_v(T - T_{sat})}\right]^{1/4} \tag{12}$$

where the characteristic length is $L_c = \sqrt{\frac{\sigma_{st}}{g(\rho_l - \rho_v)}}$, ρ_l being the liquid density, σ_{st} the surface tension between the vapor and the liquid, g the gravity, and k_v, μ_v y ρ_v the vapor thermal conductivity, viscosity and density, respectively. h'_{fg} stands for the modified latent heat that represents the energy needed to increase the liquid temperature from its bulk temperature to the saturation temperature and to transforms it into vapor:

$$h'_{fg} = C_{p,l}(T_{sat} - T_b) + i_{lv} + C_{p,v}(T - T_{sat}) \tag{13}$$

In Eq. (13), i_{lv} stands for the latent heat of evaporation while $C_{p,v}$ and $C_{p,l}$ represents the fluid specific heat for the vapor and liquid phases, respectively.

Liquid properties in Eqs. (12) and (13) are evaluated for the bulk temperature T_b, while vapor properties are taken at saturation temperature T_{sat}.

Transition boiling
At the Leidenfrost temperature T_{LDF}, calculated by [13]

$$T_{LDF} = \left(550 + 50\sqrt{V + V_{flot}} + 3(T_{sat} - T_b)\right) C_a \tag{14}$$

the vapor blanket starts to destabilize and bubble generation begins. The heat flux given by the film boiling Eq. (10) at temperature $T = T_{LDF}$ will be named q_{LDF}.

Coefficient $C_a = 1.6$ in Eq. (14) has been adjusted using laboratory quenching experiments carried out by the research group. Details of the laboratory equipment and experimental trials are explained in Sect. 2.2. Velocity V accounts for the characteristic fluid velocity of the quenching bath, which is increased in the first seconds by the immersion velocity of the pieces in the bath. V_{flot} takes into account the characteristic velocity induced by the buoyancy of the vapor bubbles in the liquid fluid:

$$V_{flot} = \sqrt{\frac{\rho_l - \rho_v}{\rho_l} g\, L} \tag{15}$$

Again, the liquid density ρ_l in Eq. (15) is evaluated at saturation temperature T_{sat}.

The heat flux dependency on the surface piece temperature T in this region has been assumed to be linear:

$$q = q_{LDF} - \left(\frac{q_{LDF} - q_{CHF}}{T_{LDF} - T_{CHF}}\right)(T_{LDF} - (T - T_{sat})) \tag{16}$$

where q_{CHF} and T_{CHF} stand for the Critical Heat Flux CHF (maximum value of the heat flux) and its corresponding temperature, detailed below.

Fully Developed Boiling

During transition boiling, the frequency of the generation of bubbles is increased, intensifying the fluid turbulence and, in consequence, the heat flux. This phenomenon has a maximum heat flux q_{CHF} [14] characterized by

$$q_{CHF} = 71987\sqrt{V + V_{flot}}(T_{sat} - T_b) \tag{17}$$

Then, as the surface piece temperature decreases (due to a lower bubble generation frequency) the heat flux decreases following [15]:

$$q = \left[1058 h_l (T - T_{sat}) \left(\rho_l (V + V_{flot}) i_{lv}\right)^{-0.7}\right]^{3.33} \tag{18}$$

where h_l represents the single-phase heat transfer coefficient, defined in Eq. (19), as indicated in [16]:

$$h_l = 0.0243 Re_l^{0.8} Pr_l^{0.4} (k_l / e) \tag{19}$$

Re_l and Pr_l are the liquid Reynolds and Prandtl number evaluated at the liquid bulk temperature T_b. The Reynolds number is based on the fluid velocity V and the characteristic length of the piece perpendicular to the incoming flow e. Temperature T_{CHF} for the maximum heat flux is obtained as the temperature when Eq. (18) reaches the maximum heat transfer value q_{CHF} given by Eq. (17).

Partial Boiling

When the contribution of the nucleate boiling heat flux becomes of the same order as the single phase forced convection, a smooth transition between this two regimes, defined by (18) and (19) respectively, is expected. The Partial Boiling regime is delimited by 2 points, called FDB (Fully Developed Boiling) and ONB (Onset of Nucleate Boiling) [17].

The estimation of the FDB point follows the following relationship:

$$q_{FDB} = 1.4 q_D \tag{20}$$

where the heat flux q_D is the intersection of the single-phase correlation (19) with the fully developed boiling Eq. (18) and is determined iteratively by Eq. (21):

$$q_D = \left[1058 (q_D - h_l (T_{sat} - T_b)) \left(\rho_l (V + V_{flot}) i_{lv}\right)^{-0.7}\right]^{3.33} \tag{21}$$

Once q_{FDB} is obtained, T_{FDB} is determined by (18). The temperature T_{ONB}, which marks the end of the Partial Boiling regime, is obtained as follows [17]:

$$T_{ONB} = T_{sat} + \frac{4\sigma_{st} T_{sat} h_l}{i_{lv} k_l \rho_v} \left[1 + \sqrt{1 + \frac{k_l i_{lv} (T_{sat} - T_b) \rho_v}{2\sigma_{st} T_{sat} h_l}}\right] \tag{22}$$

with the properties of the liquid evaluated at T_b. Its associated heat flux q_{ONB} is obtained from the corresponding single-phase correlation of Eq. (19) for this temperature.

Finally the heat flux for this region is obtained from the following equation:

$$q = q_{ONB} + \frac{q_{FDB} - q_{ONB}}{(T_{FDB} - T_{sat})^m - (T_{ONB} - T_{sat})^m} \cdot \\ \cdot \left((T - T_{sat})^m - (T_{ONB} - T_{sat})^m\right) \tag{23}$$

where m is obtained, for each point, by

$$m = 1 + \frac{2.33}{q_{FDB} - q_{ONB}}(q - q_{ONB}) \tag{24}$$

starting with $m = 3.33$ at T_{FDB} and ending with $m = 1$ at T_{ONB}.

Single phase

Finally, the bubble generation ends so this regime is characterized by the forced convection between the piece and the liquid. For temperatures below T_{ONB}, the Dittus-Boelter correlation (19) is used. However, for low temperatures (lower than $T_b + 25K$) Eq. (19) over-predicts significatively the heat transfer. Thus, CFD techniques have been used to adjust (by least-squares techniques) a new convective heat transfer that adapts to different agitation velocities and takes into account the variation of the fluid viscosity with temperature:

$$h_{l1} = 0.097 Re_l^{0.612} Pr_l^{0.23} \left(\frac{\mu_l}{\mu_{sat}}\right)^{0.175} \left(\frac{k_l}{e}\right) \tag{25}$$

Summarizing, the proposed expressions in Sect. 2.1.2 define the temperature dependent heat flux $q(T)$ to the quenching fluid. This function has been programmed using *MATLAB*[2] and it is shown in Fig. 3 for the specific values of the industrial quenching process (mainly agitation velocity V and fluid temperature T_b) analyzed in this work.

2.2 Thermal Model Validation

The corresponding heat flux to air defined in Sect. 2.1.1 and the previous temperature dependent heat flux $q(T)$ from Sect. 2.1.2 were exported to software COMSOL-Multiphysics[3] where the complete thermal model, defined by 1 and 2, has been

[2]https://www.mathworks.com/.

[3]https://www.comsol.com/.

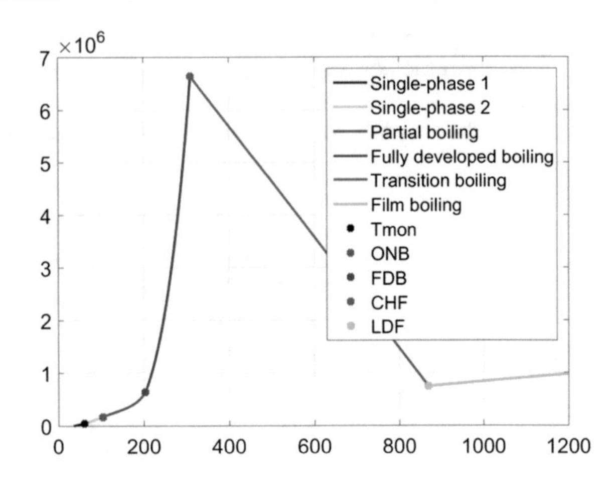

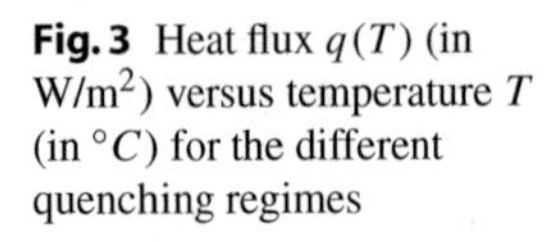
Fig. 3 Heat flux $q(T)$ (in W/m^2) versus temperature T (in $^\circ C$) for the different quenching regimes

Table 1 Fluid conditions for the thermal model validation

T_b (K)	V_1 (m/s)	V_2 (m/s)	V_3 (m/s)
303	0.75	0.55	0.25
343	0.75	0.55	0.25

numerically solved (for a material without metallurgical transformations) and validated by comparison with experiments on a standard laboratory probe.

The cylindrical test probes (of 60 mm of length and 12.5 mm of diameter) comply with the international standard ISO 9950 and the American standards ASTM D 6200-01 and ASTM D 6482-99. A $k-$type thermocouple, localized at the center of the probe, is connected to a data acquisition unit that saves temperature data each 0.01 s. The probe material is Inconel 600, which does not undergo any change in its microstructure when cooled. Consequently, the model has been solved for $Q = 0$ while the dependency of the Inconel properties ρ, k and Cp with temperature have been extracted from the technical literature.

These experiments were carried out under different fluid conditions, shown in Table 1. Comparison of the experimental results, that is, the evolution of the temperature at the center of the probe $T(t)$ as well as its cooling velocity $\frac{\partial T}{\partial t}$, and the corresponding numerical results was done. The averaged relative errors between the numerical and experimental results were lower than 10%. An example is shown in Fig. 4, where experimental profiles of T versus time t and cooling velocity $\frac{\partial T}{\partial t}$ versus T are depicted in red color, while numerical results given by the model for the same process conditions are indicated in blue color.

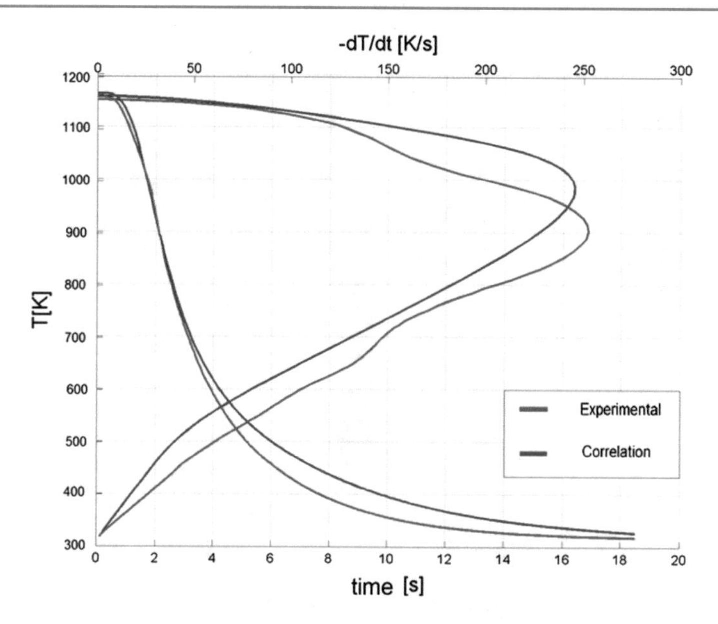

Fig. 4 Thermal model validation against experimental data for T_b =303 K and $V = 0.55$ m/s

2.3 Metallurgical Model

The evolution of the microstructural phases dominated by carbon diffusion process (ferrite, f, pearlite, p, and bainite, b), have been modeled by an Avrami type equation [18] and [19]:

$$X_i = X_\gamma X_{i,max} \left[1 - exp\left(-b\left(\frac{d_\gamma}{d_{\gamma,ref}}\right)^m t^n\right)\right] \tag{26}$$

where X_i stands for the proportion of micro-constituent i, $X_{i,max}$ is the maximum proportion of micro-constituent i at a given temperature and X_γ the proportion of austenite at the beginning of the transformation. To obtain the material parameters b and n, a TTT (Time Temperature Transformations) diagram of the $25M6$ steel is used due to the similar chemical composition. Parameter $d_{\gamma,ref}$ in Eq. (26) is the reference austenitic grain size of the TTT diagram while d_γ is the grain size for the processed material, equal in this case to 6.5 ASTM for the industrial spindles.

As the industrial process is a continuous cooling, the additive rule of Scheil [10] is used to discretize it in small isothermal intervals. The beginning of the transformations is determined by

$$\sum \frac{\Delta t_i}{\tau_i\left(\frac{d_\gamma}{d_{\gamma,ref}}\right)} \geq 1 \tag{27}$$

where τ_i stands for the incubation time for a specific micro-constituent for each temperature, given by the TTT diagram.

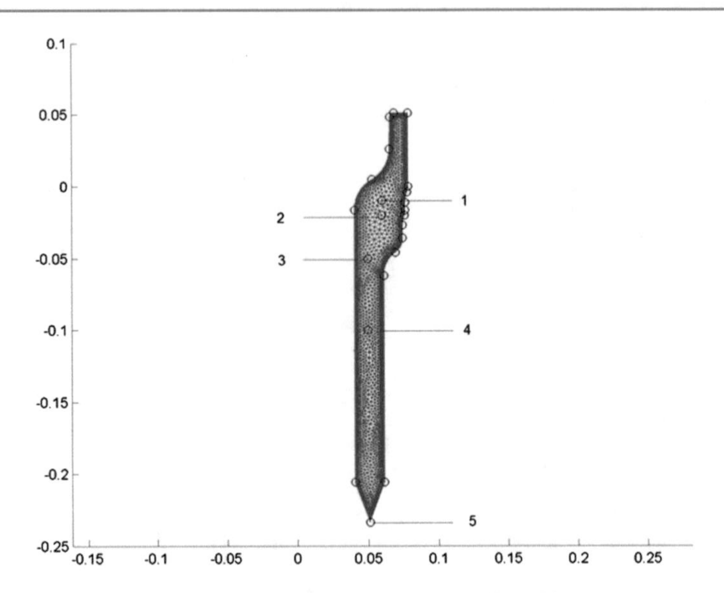

Fig. 5 Computational mesh for the spindle. Dimensions are indicated in m

The non-diffusion martensitic transformation is modeled using the Koïstinen-Marburger law [20]:

$$X_m = \left(1 - X_f - X_p - X_b\right)\left\{1 - exp\left(\beta[M_s - T]^+\right)\right\} \tag{28}$$

β being a material parameter and M_s the temperature for the beginning of the martensitic transformation [21].

2.4 Implementation

The complete thermal-metallurgical transient model defined in Sect. 2 has been numerically solved through Finite Element Methods (FEM) in COMSOL-Multiphysics software coupled to MATLAB. The spindle geometry has been simplified to a $2D$ axisymmetric domain, discretized with a mesh of 3390 nodes and 6176 triangular elements, shown in Fig. 5.

In order to analyze the thermal and the metallurgical evolution, 5 selected points, shown in Fig. 5, were selected.

For the calculations, a workstation of 128 Gb of RAM with two Intel-Xeon processors (6 nodes and 1.8 GHz) was used. Computational times of each case took around 15 h.

The initial conditions for the computational domain were a completed austenized piece ($X_\gamma = 1$) and homogeneous temperature equal to $T = 1163$ K.

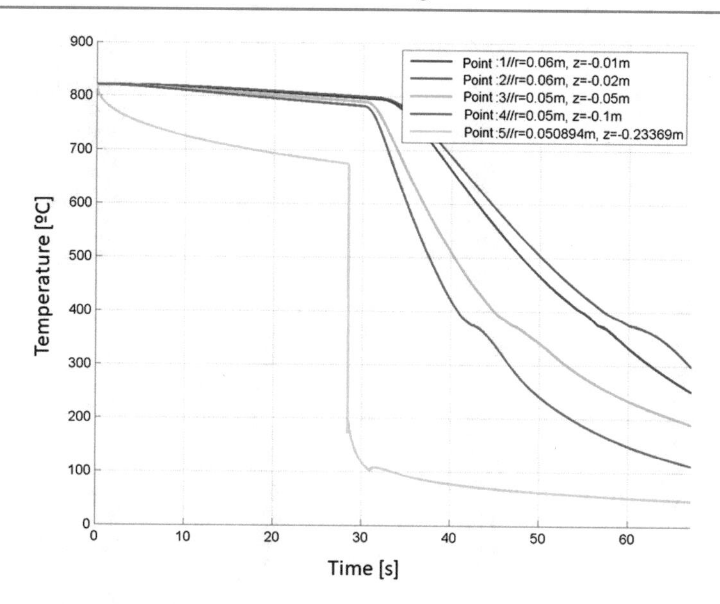

Fig. 6 Thermal evolution $T(t)$ at the 5 monitored points

3 Results

To illustrate the thermal evolution of the piece, we monitored the temperature at the 5 points previously indicated in Fig. 5 for the industrial quenching parameters of $T_b = 303$ K and $V = 0.35$ m/s. Analysing Fig. 6, we can observe the change of slope at $t = 27$ s associated with the increase in the cooling velocity due to the end of the cooling in air and the start of the quenching stage. On the other hand, the heat released by the martensitic transformation produces a reheating effect around the transformation temperature of 360 °C (small bumps in Fig. 6).

The final metallurgical phases proportions predicted by the model are shown in Fig. 7. The maximum bainite content (up to a 5.8%) is obtained at the thickest region of the piece. This bainite distribution is in agreement with the metallurgical studies (micrographs analysis) carried out by the company on random spindles manufactured under the evaluated quenching operating parameters.

4 Conclusions

A numerical thermal-metallurgical model, which simplifies the heat transfer phenomena between the piece surface and the fluid, is presented. The model uses a set of literature correlations as well as some CFD adjusted correlations and it has been adapted to the industrial quenching process parameters.

Before its application, the thermal model was validated with experimental data on cylindrical probes at laboratory scale under different process parameters. The

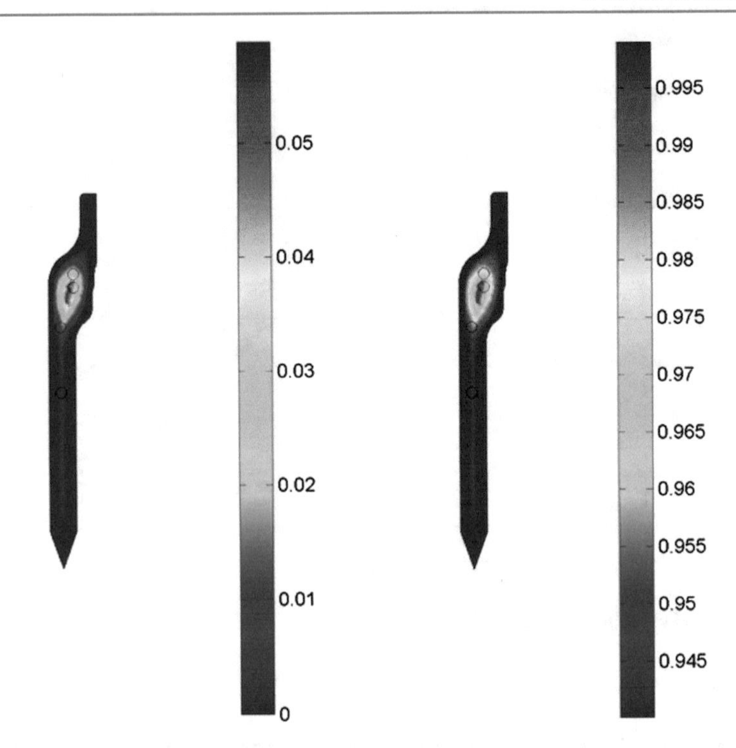

Fig. 7 Final metallurgical phases proportion (left: bainite, right: martensite) predicted by the model

thermal model is able to describe the different heat transfer regimes that occur with relative accuracy (relative errors are below 10%).

The final model is capable of solving the metallurgical evolution during the complete industrial process for the different process parameters (e.g., fluid temperature, fluid agitation velocity, transportation time, steel chemical composition and furnace homogenization temperature) at a reduced computational time. In consequence, the model allows the feasibility evaluation of different industrial parameters to assess the suitability (in terms of the final mechanical properties of the pieces) of more efficient arrangements of the production line.

Acknowledgements This research was supported by the Centre for the Development of Industrial Technology (CDTI), Ministry of Science, Innovation and Universities of Spain (with European Regional Development Fund) in collaboration with company CIE-Galfor S.A., under grant IDI-20170068 (Project acronym: TEINEXT).

References

1. Dhir, V.K.: Boiling heat transfer. Ann. Rev. Fluid Mechan. **30**, 365–401 (1998)
2. Nukiyama, S.: The maximum and minimum values of the heat Q transmitted from metal to boiling water under atmospheric pressure. Int. J. Heat Mass Trans. **9**, 1419–1433 (1966)

3. Yeoh, G.H., Tu, J.Y.: Population balance modelling for bubbly flows with heat and mass transfer. Chem. Eng. Sci. **59**, 3125–3139 (2004)
4. Kocamustafaogullari, G., Ishii, M.: Foundation of the interracial area transport equation and its closure relations. Int. J. Heat Mass Trans. **38**(3), 481–493 (1995)
5. Kocamustafaogullari, G., Ishii, M.: Interfacial area and nucleation site density in boiling systems. Int. J. Heat Mass Trans. **26**(9), 1377–1387 (1983)
6. Meduri, P.K., Warrier, G.R., Dhir, V.K.: Wall heat flux partitioning during subcooled forced flow film boiling of water on a vertical surface. Int. J. Heat Mass Trans. **52**, 3534–3546 (2009)
7. Bromley, L.A., Leroy, N.R., Robbers, J.A.: Heat transfer in forced convection film boiling. Ind. Eng. Chem. **45**(12), 2639–2646 (1953)
8. Campbell, P.C.: Application of microstructural engineering to the controlled cooling of steel wire rod. Ph.D. thesis, The University of British Columbia (1989)
9. Huiping, L., Guoqun, Z., Lianfang, H., Yue, M.: Solution of non-linear thermal transient problems by a new adaptive time-step method in quenching process. Appl. Math. Modell. **33**, 329–342 (2009)
10. Nobari, A.H., Serajzadeh, S.: Modeling of heat transfer during controlled cooling in hot rod rolling of carbon steels. Appl. Therm. Eng. **31**, 487–492 (2011)
11. White, F.M.: Heat and Mass Transfer. University of Rhode Island, Addison-Wesley. (1988)
12. Meduri, P.K.: Wall heat flux partitioning during subcooled flow film boiling of water on a vertical surface. Ph.D. thesis, University of California, Los Angeles (2007)
13. Drucker, M., Dhir, V.K.: Effects of high temperature and flow blockage on the reflood behavior of a 4-rod boundle. Technical report, EPRI (1981)
14. Inasaka, F., Nariai, H.: Evaluation of subcooled critical heat flux correlations for tubes with and without internal twisted tapes. Nucl. Eng. Des. **163**, 225–239 (1996)
15. Prodanovic, V., Fraser, D., Salcudean, M.: On the transition from partial to fully developed subcooled flow boiling. Int. J. Heat Mass Trans. **45**, 4727–4738 (2002)
16. Gungor, K.E., Winterton, R.H.S.: A general correlation for flow boiling in tubes and annuli. Int. J. Heat Mass Trans. **29**(3), 351–358 (1986)
17. Kandlikar, S.G.: Heat transfer characteristics in partial boiling, fully developed boiling, and significant void flow regions of subcooled flow boiling. J. Heat Trans. **120**, 395–401 (1998)
18. Viéitez, I., L.-Cancelos, R., Martín, E., Varas, F.: Predictive model of wire rod cooling. In: Proceedings of the 28th ASM Heat Treating Society Conference, pp. 518–524 (2015)
19. Fachinotti, V.D., Cardona, A., Anca, A.A.: Solid-state microstructure evolution in steels. In: Mecánica Computacional, XXIV. MECOM (2005)
20. Koïstinen, D.P., Marburger, R.E.: A general equation prescribing the extent of the austenite-martensite transformation in pure iron-carbon alloys and plain carbon steels. Acta Metallurgica **7**(1), 59–60 (1959)
21. Pero-Sanz, J.A. Aceros. Metalurgia física, Selección y Diseño. Dossat (2004)

Single Particle Models for the Numerical Simulation of Lithium-Ion Cells

Alfredo Ríos-Alborés and Jerónimo Rodríguez

Abstract

In the design of Battery Management Systems (BMS) for a lithium-ion cell, it is crucial to accurately simulate the device in real time using mathematical models. Often, Equivalent Circuit Models (ECM) are used to this end, due to their simplicity and efficiency. However, they are purely phenomenological (their parameters are fitted to emulate empirical data) and their internal variables lack physical meaning. On the other hand, the most popular physics-based electrochemical model in the literature, the pseudo-two-dimensional (P2D) model, presents a high computational cost. In this paper, we review the single particle model (SPM), a physics-based model of reduced complexity that is suitable for real-time applications.

1 Introduction

In the last decades, there has been an increasing interest in the development and improvement of electric energy storage devices. The electrochemical batteries based on lithium-ion chemistry present good properties, such as high energy and power density, long life expectancy, low self-discharge rate, non-memory effect, among others [8]. Its advantages, compared to other chemistries, make this technology the preferred candidate for electrical vehicles [14]. However, lithium-ion cells are

A. Ríos-Alborés (✉) · J. Rodríguez
Departamento de Matemática Aplicada, Universidade de Santiago de Compostela, 15782 Santiago de Compostela, Spain
e-mail: Jeronimo.Rodriguez@usc.es

ITMATI, Campus Sur, 15706 Santiago de Compostela, Spain
e-mail: alfredo.rios.albores@usc.es

J. Rodríguez
IMAT, Universidade de Santiago de Compostela, 15706 Santiago de Compostela, Spain

P. Quintela Estévez et al. (eds.), *Advances on Links Between Mathematics and Industry*, SxI - Springer for Innovation / SxI - Springer per l'Innovazione 15,
https://doi.org/10.1007/978-3-030-59223-3_6

sensitive to inappropriate use conditions [3,4]. Hence, for safety reasons and to improve their performance, it is very important to estimate the cell's state.

For real-time applications, like BMS, a mathematical model is necessary, as a virtual cell. Often, an ECM is implemented [25]. In these models, which are purely phenomenological, their parameters are fitted to reproduce empirical measures. In consequence, the internal variables lack true physical meaning. Modern BMS pretend to implement advanced features, such as optimal/fast charge protocols, cell degradation estimation, and internal cell states monitoring [17,18]. For such features, physics-based models are better suited than equivalent models.

In the literature, the most popular validated physics-based model is the P2D model, firstly proposed in [9]. From a mathematical point of view, the P2D model can be formulated as a non-linear partial differential equations (PDEs) system, of parabolic and elliptic equations, all of them coupled by non-linear algebraic equations, in particular, Butler–Volmer kinetics equations [12]. Due to its complexity, its computational cost is prohibitive for real-time applications. Different approaches are found in the literature to simplify this model. For example, order reduction techniques [5,6,10,11] or just simplifying assumptions [30,31]. Among the latter, the SPM is one of the most popular choices. Its formulation is deduced from the P2D model under the main assumption that the intercalation/deintercalation reaction flux across each electrode is homogeneously distributed.

The first examples of SPM considered that the electric potentials and electrolyte physics in the cell are negligible under low C-rate current profiles [22]. In [28], a SPM was compared with a P2D model. The SPM performance was acceptable for low C-rates protocols, up to $1C$. In other works, lithium-ion distribution is modeled in terms of average concentrations in the solid particles, obtaining linear ordinary differential equations, and allowing for a readily implementation of linear filter techniques, such as the Kalman filter [7,29]. Different approaches have been proposed to enhance the SP model for its use under higher C-rates currents [13,19,26]. Furthermore, it can be extended to capture thermal dynamics [2,23] and to estimate cell degradation over-time [15,27], using meaningful internal electrochemical quantities of the cells.

In this paper, we review the single particle model with electrolyte dynamics (SPMe) [16,21]. We focus on its derivation from the P2D model, stating and describing the main physical simplifying assumptions. The SPMe is numerically solved using the finite element library FEniCS [1]. The results are compared to those provided by the P2D model in terms of applicability range, accuracy, and computational cost.

2 The P2D Model

The equations of physics-based models for lithium-ion cells can be naturally deduced applying conservation laws for mass and charge at the three-dimensional particle scale or "micro-scale." Then, using volume-averaging techniques, and under some simplifying assumptions, one can obtain the P2D model equations. Further details and gentle explanations of all this can be found in [24].

Table 1 Nomenclature

Symbol	Units	Name and description
Greek symbols		
$\alpha, 1-\alpha$	1	Asymmetric charge transfer coefficients
$\varepsilon_e, \varepsilon_s$	1	Volume fraction of the electrolyte and the solid phases
κ	S/m	Ionic conductivity of the electrolyte
Ω		Macroscale cell domain
$\Omega^-, \Omega^+, \Omega^{\mathrm{o}}$		Macroscale subdomains: negative and positive electrode, separator
Ω_{ap}		Microscale solid phase domain
ϕ_e, ϕ_s	V	Electric potential in the electrolyte and solid phase
$\sigma^{\pm}$	S/m	Electric conductivity of the solid phase
θ	K	Temperature
Latin symbols		
a_s	$\mathrm{m}^2/\mathrm{m}^3$	Surface area density of solid particles
brugg	1	Bruggeman coefficient
c_e	$\mathrm{mol}/\mathrm{m}^3$	Lithium salt concentration in the electrolyte
$c_s^{\pm}$	$\mathrm{mol}/\mathrm{m}^3$	Intercalated lithium concentration in the solid particles
$c_{s,\max}^{\pm}$	$\mathrm{mol}/\mathrm{m}^3$	Maximum concentration of intercalated lithium
$D_s^{\pm}, D_e$	m^2/s	Solid phase and electrolyte diffusion coefficients
F	C/mol	Faraday's constant
i	A	Current intensity
$j_{Li}^{\pm}$	$\mathrm{mol}/\mathrm{m}^2\mathrm{s}$	Lithium intercalation-deintercalation reaction flux at $\Omega^{\pm}$
$k_0^{\pm}$	$\frac{\mathrm{mol}^{\frac{-1}{2}}\,\mathrm{m}^{\frac{5}{2}}}{\mathrm{s}}$	Effective rate constant of the in-deintercalation reaction
L^-, L^+, L^{o}	m	Eectrodes and separator thickness
L	m	Cell thickness, $L := L^- + L^+ + L^{\mathrm{o}}$
r	m	Microscale radial space variable
r_{SEI}	m	SEI layer radial thickness
R	J/K mol	Ideal gas constant
R_{col}	Ω	Total current collectors resistance
$R_s^{\pm}$	m	Average solid particles radius in each electrode
S	m^2	Current collectors area
t_+^o	1	Ionic transfer number of the electrolyte
$U_{\mathrm{ocp}}^{\pm}$	V	OCP of the positive/negative electrode solid material
V	V	Cell voltage
x	m	Macroscale space variable
$x_{0\%}, x_{100\%}, y_{0\%}, y_{100\%}$	1	Nominal stoichiometry window
$z_{\mathrm{film}}^{\pm}$	$\Omega\mathrm{m}^2$	Resistance of the film on the solid particles surface

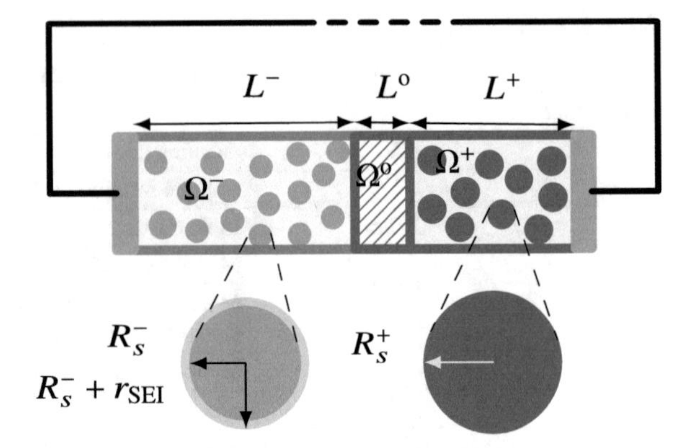

Fig. 1 Sketch of a cell. On the left (resp. right), the negative electrode, Ω^- (resp. positive electrode, Ω^+). In the middle, the separator, Ω^o, which is an electric insulator. At the micro-scale level, the electrode's solid particles are assumed to be spheres. Usually, the SEI layer is formed on the surface of the particles in Ω^-. Modified with permission from [12]

For completeness, we state a P2D model equations. We assume that the cell temperature is spatially homogeneous and known at every time, and neglect cell degradation over-time mechanisms. We use a second-order elliptic formulation for the electric potentials, which is a better-suited formulation for the model resolution using the finite element method. For brevity, we avoid presenting symbols, which are summarized in Table 1.

It is assumed that all the main processes occur in the direction perpendicular to the cell. Therefore, only the cell thickness is accounted as a spatial dimension and the domain of the problem, representing the device, is modeled as an 1D-domain, $\Omega = (0, L)$. Within, we distinguish the negative electrode, the separator, and the positive electrode as subdomains, denoted, respectively, as $\Omega^- = \left(0, L^-\right)$, $\Omega^o = \left(L^-, L^- + L^o\right)$ and $\Omega^+ = \left(L^- + L^o, L\right)$. Furthermore, to model the intercalated lithium diffusion in the solid reactive particles at each electrode, we need to introduce the 2D microscale domain $\Omega_{ap} = \left(\left(0, R_s^-\right) \times \Omega^-\right) \cup \left(\left(0, R_s^+\right) \times \Omega^+\right)$. A simple sketch of a cell is represented in Fig. 1. The model equations are stated as follows.

2.1 Butler–Volmer Equation

At every $x \in \Omega^\pm$, the intercalation/deintercalation reaction flux on the surface of the solid particle is modeled using Butler–Volmer kinetics, namely,

$$j_{Li}^{\pm}(x,t) = \frac{i_0^{\pm}(x,t)}{F}\left(\exp\left(\frac{\left(1-\alpha^{\pm}\right)F}{R\theta}\eta^{\pm}(x,t)\right) - \exp\left(-\frac{\alpha^{\pm}F}{R\theta}\eta^{\pm}(x,t)\right)\right), \tag{1}$$

where

$$i_0^{\pm}(x,t) = Fk_0^{\pm}\left(c_e(x,t)\left(c_{s,\max}^{\pm} - c_s^{\pm}\left(R_s^{\pm},x,t\right)\right)\right)^{1-\alpha^{\pm}}\left(c_s^{\pm}\left(R_s^{\pm},x,t\right)\right)^{\alpha^{\pm}}, \tag{2}$$

$$\eta^{\pm}(x,t) = \phi_s^{\pm}(x,t) - \phi_e(x,t) - U_{\text{ocp}}^{\pm}\left(\frac{c_s^{\pm}(R_s^{\pm},x,t)}{c_{s,\max}^{\pm}}\right) - F z_{\text{film}}^{\pm} j_{Li}^{\pm}(x,t). \quad (3)$$

We further define

$$j(x,t) := \begin{cases} j_{Li}^{-}(x,t), & x \in \Omega^{-}, \\ 0, & x \in \Omega^{o}, \\ j_{Li}^{+}(x,t), & x \in \Omega^{+}. \end{cases} \quad (4)$$

2.2 Intercalated Lithium Concentration in the Solid Particles

The distribution of lithium inside the particles is assumed to respond only to diffusion effects. The corresponding parabolic PDE is defined in the micro-scale 2D-domain, namely,

$$\frac{\partial c_s}{\partial t}(r,x,t) - \frac{1}{r^2}\frac{\partial}{\partial r}\left(D_s r^2 \frac{\partial c_s}{\partial r}(r,x,t)\right) = 0, \text{ in } \Omega_{\text{ap}}, \quad (5)$$

$$\frac{\partial c_s}{\partial r}(0,x,t) = 0, \quad \forall (0,x) \in \Omega_{\text{ap}}, \quad (6)$$

$$-D_s \frac{\partial c_s^{\pm}}{\partial r}\left(R_s^{\pm},x,t\right) = j_{Li}^{\pm}(x,t), \quad \forall \left(R_s^{\pm},x\right) \in \{R_s^{\pm}\} \times \Omega^{\pm}, \quad (7)$$

where $c_s^{\pm}(r,x,t) := c_s(r,x,t)$ in $\left(0,R_s^{\pm}\right) \times \Omega^{\pm}$. Notice that only the boundary condition depends on $x \in \Omega^{\pm}$. Initial conditions, $c_s^{\pm}(r,x,0)$, have to be given.

2.3 Lithium Salt Concentration in the Electrolyte

For the lithium salt distribution in the electrolyte, the model accounts for the diffusion effects, the electroneutrality condition of the medium and the reaction flux as a source term,

$$\frac{\partial(\varepsilon_e c_e)}{\partial t}(x,t) - \frac{\partial}{\partial x}\left(D_e^{\text{eff}} \frac{\partial c_e}{\partial x}(x,t)\right) = \left(1 - t_{+}^{o}\right) a_s j(x,t), \text{ in } \Omega, \quad (8)$$

$$\frac{\partial c_e}{\partial x}(0,t) = \frac{\partial c_e}{\partial x}(L,t) = 0, \quad (9)$$

where $D_e^{\text{eff}}(x,t) \equiv D_e(\theta)\,\varepsilon_e(x,t)^{\text{brugg}}$. An initial condition, $c_e(x,0)$, has to be given.

2.4 Electric Potential in the Electrolyte

The electric potential in the electrolyte spatially varies due to the lithium salt gradient effects and due to the electrochemical reaction flux distribution across the cell,

$$\frac{\partial}{\partial x}\left(\kappa^{\text{eff}}\frac{\partial \phi_e}{\partial x}(x,t)\right)+\frac{\partial}{\partial x}\left(\kappa_D^{\text{eff}}\frac{\partial \ln c_e}{\partial x}(x,t)\right)=-Fa_s j(x,t), \text{ in } \Omega, \tag{10}$$

$$\frac{\partial \phi_e}{\partial x}(0,t)=\frac{\partial \phi_e}{\partial x}(L,t)=0, \tag{11}$$

where $\kappa^{\text{eff}}(x,t)=\kappa(c_e,\theta)\,\varepsilon_e(x)^{\text{brugg}}$, $\kappa_D^{\text{eff}}(x,t)=\kappa_D(x,t)\,\varepsilon_e(x)^{\text{brugg}}$ and $\kappa_D(x,t):=\frac{2\kappa(c_e,\theta)R\theta(t_+^o-1)}{F}$.

2.5 Electric Potential in the Solid Phase

In the solid phase, the electric potential spatially varies due to the reaction flux distribution across each electrode,

$$\frac{\partial}{\partial x}\left(-\sigma^{\text{eff}}\frac{\partial \phi_s}{\partial x}(x,t)\right)=-Fa_s j(x,t), \text{ in } \Omega^-\cup\Omega^+, \tag{12}$$

$$\frac{\partial \phi_s^-}{\partial x}\left(L^-,t\right)=\frac{\partial \phi_s^+}{\partial x}\left(L^-+L^{\text{o}},t\right)=0, \tag{13}$$

$$-\sigma^{\text{eff}}\frac{\partial \phi_s^-}{\partial x}(0,t)=-\sigma^{\text{eff}}\frac{\partial \phi_s^+}{\partial x}(L,t)=\frac{i(t)}{S}, \tag{14}$$

where $\sigma^{\text{eff}}(x,t)=\sigma(\theta)\,\varepsilon_s(x)^{\text{brugg}}$. We are assuming that the current signal, $i(t)$, is given, so that the PDE is stated using Neumann boundary conditions.

2.6 Gauge Conditions on Potentials

Notice that we cannot expect uniqueness of solution for this model. Indeed, disregarding the Butler–Volmer equation, both potentials are always affected by a derivative. Moreover, the dependence on the potentials in the Butler–Volmer equation is through the difference $\phi_s(x,t)-\phi_e(x,t)$. In consequence, if we find a solution to the system, by adding a constant to both potentials (the same for both), we would get a different solution. For that reason, it is necessary to impose a gauge condition only on one of them. In this study, we propose the condition

$$\int_\Omega \phi_e(x,t)\ \mathrm{d}x=0. \tag{15}$$

3 Derivation of the SPM

We will now derive the SPM from the P2D equations presented in the previous section. A similar derivation process has been addressed in [21].

We consider that, initially, the cell is in rest and in steady-state. Hence, the spatial distribution of intercalated lithium in the solid phase at each electrode is approximately homogeneous, $c_s^{\pm}(r, x, 0) = c_{s,0}^{\pm}$, with $c_{s,0}^{\pm} \in \mathbb{R}^{+}$. And the same is true for the lithium salt in the electrolyte across the entire cell, $c_e(x, 0) = c_{e,0}$, with $c_{e,0} \in \mathbb{R}^{+}$. Then, we assume

- A_1 Some cell material properties are constant per subdomain. In particular: $\varepsilon_e(x)$, $\varepsilon_s(x)$, $R_s(x)$, and $\kappa(c_e, \theta)$.
- A_2 The electrochemical reaction flux distribution of the intercalation/deintercalation process is homogeneous across each electrode. Namely, $j_{Li}^{\pm}(x, t) \equiv j_{Li}^{\pm}(t)$, in $\Omega^{\pm}$.
- A_3 $\alpha^{\pm} = \left(1 - \alpha^{\pm}\right) \equiv \alpha = 1/2$.

Remark 1 Assumption A_3 is not strictly necessary to derive a single particle model. Nevertheless, it simplifies the resolution of the model (see *Remark* 2). In the literature, A_3 is assumed frequently, even when dealing with the P2D model.

Next, we apply assumptions $A_1 - A_3$ to the P2D model (1–15). We will refer to figures of numerical result to illustrate qualitative properties of the new model equations, even though the technical details of those numerical experiments will not be given until the next section.

3.1 Butler–Volmer Equation

Now, (1 – 3) become

$$j_{Li}^{\pm}(t) = \frac{1}{F} i_{0,\text{avg}}^{\pm}(t) \left(\exp\left(\frac{\alpha F}{R\theta} \eta^{\pm}(t) \right) - \exp\left(-\frac{\alpha F}{R\theta} \eta^{\pm}(t) \right) \right), \text{ in } \Omega^{\pm}, \quad (16)$$

$$i_{0,\text{avg}}^{\pm}(t) := \frac{1}{L^{\pm}} \int_{\Omega^{\pm}} F k_0^{\pm} \left(\left(c_{s,\max}^{\pm} - c_s^{\pm}\left(R_s^{\pm}, t\right)\right) c_e(x, t)\, c_s^{\pm}\left(R_s^{\pm}, t\right)\right)^{\alpha} \, \mathrm{d}x, \quad (17)$$

$$\eta^{\pm}(t) = \phi_s^{\pm}(x, t) - \phi_e^{\pm}(x, t) - U_{\text{ocp}}^{\pm}\left(\frac{c_s^{\pm}\left(R_s^{\pm}, t\right)}{c_{s,\max}^{\pm}} \right) - F z_{\text{film}}^{\pm} j_{Li}^{\pm}(t). \quad (18)$$

In each electrode, we have approximated the exchange current density $i_0^{\pm}(x, t)$ in (2) by its average, $i_{0,\text{avg}}^{\pm}(t)$, and assumed that the overpotential $\eta^{\pm}(x, t) \approx \eta^{\pm}(t)$. In practice, $k_0^{\pm}$ is usually given as a constant value per electrode. In that case, to compute $i_{0,\text{avg}}^{\pm}(t)$, it would be enough to average the lithium salt distribution function $c_e(x,t)^{\alpha}$ in $\Omega^{\pm}$. Notice that, to all this to be physically accurate, $c_e(x,t)$ and the difference $\phi_s(x,t) - \phi_e(x,t)$ have to be *sufficiently homogeneous* across the electrodes.

Integrating (12) across each electrode, applying boundary conditions (13 – 14), we obtain an explicit linear expression for $j(x,t)$ as a linear function of the current,

$$j(x,t) = \begin{cases} j_{Li}^{-}(t) = \frac{i(t)}{FSL^{-}a_s^{-}}, \text{ in } \Omega^{-}, \\ 0, \text{ in } \Omega^{o}, \\ j_{Li}^{+}(t) = -\frac{i(t)}{FSL^{+}a_s^{+}}, \text{ in } \Omega^{+}. \end{cases} \tag{19}$$

Furthermore, (16) is analytically invertible, and its left-hand side is given by (19). Then,

$$\eta^{\pm}(t) = \frac{R\theta}{\alpha F}\sinh^{-1}\left(\frac{j_{Li}^{\pm}(t)F}{2i_{0,\text{avg}}^{\pm}(t)}\right) = \frac{R\theta}{\alpha F}\sinh^{-1}\left(\mp\frac{i(t)}{2i_{0,\text{avg}}^{\pm}(t)SL^{\pm}a_s^{\pm}}\right). \tag{20}$$

Remark 2 Notice that we have used A_3 to obtain this closed form for $\eta^{\pm}(t)$. The values of coefficients $\alpha^{\pm}$ could be, eventually, different. This would require solving two non-linear equations, similar to (16), to obtain the overpotentials $\eta^{\pm}(t)$.

3.2 Intercalated Lithium Concentration in the Solid Particles

The PDE (5 – 7) is no longer x-dependent, because $j_{Li}^{\pm}(t)$ is homogeneous across $\Omega^{\pm}$. Moreover, we are assuming that $\varepsilon_s(x)$ and $R_s(x)$ are constant per subdomain (A_1), i.e., the distribution and size of the solid particles are homogeneous and constant across each electrode. As a consequence, every solid particle in $\Omega^{\pm}$ will behave the same and, then, $c_s^{\pm}(r,x,t) \equiv c_s^{\pm}(r,t)$. It will be enough to solve the following two linear parabolic PDEs,

$$\frac{\partial c_s^{\pm}}{\partial t}(r,t) - \frac{1}{r^2}\frac{\partial}{\partial r}\left(D_s r^2 \frac{\partial c_s^{\pm}}{\partial r}(r,t)\right) = 0, \quad r \in \left(0, R_s^{\pm}\right), \tag{21}$$

$$\frac{\partial c_s^{\pm}}{\partial r}(0,t) = 0, \tag{22}$$

$$-D_s\frac{\partial c_s^{\pm}}{\partial r}\left(R_s^{\pm},t\right) = \mp\frac{i(t)}{FSL^{\pm}a_s^{\pm}}. \tag{23}$$

Notice that the two-dimensional micro-scale domain of the P2D model has been simplified to a one-dimensional domain. In each electrode, the intercalated lithium distribution profile of the single particle, modeled by (21 – 23), will be an average of the heterogeneous distribution of solid particles across the electrode in the P2D model, as it is shown in Figs. 2 and 3.

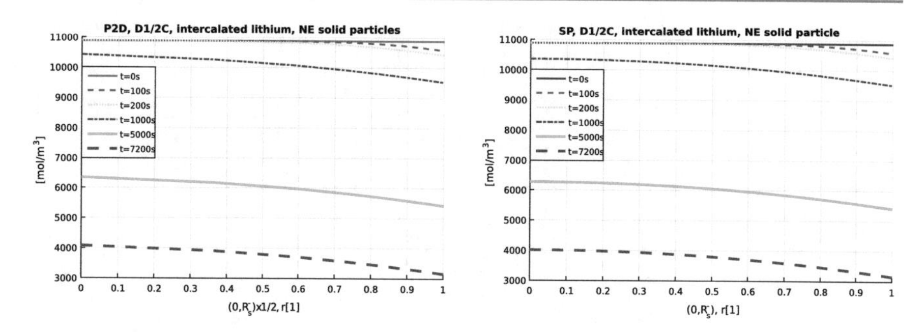

Fig. 2 Discharge under constant $0.5C$-rate. Lithium distribution within particles of the negative electrode, at the middle of the electrode ($x = 1/2$) for the P2D model (left), and for the SP model (right)

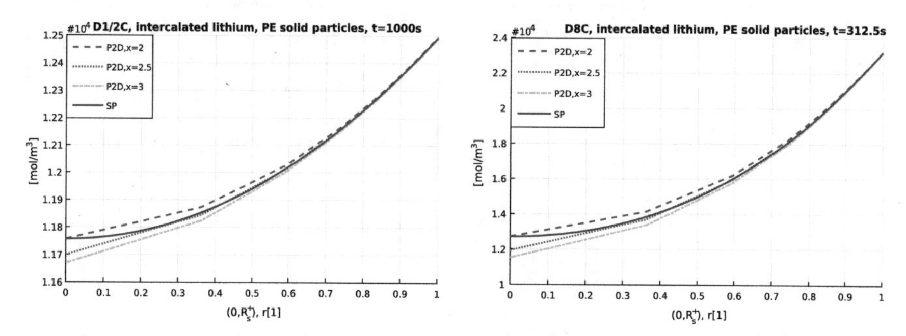

Fig. 3 Discharge under constant $0.5C$-rate (left) and $8C$-rate (right). Lithium concentration distribution inside the solid particles of the positive electrode. For the P2D model, we are representing values inside the particles at the end sides and at the middle of the electrode

3.3 Lithium Salt Concentration in the Electrolyte

Under assumptions A_1 and A_2, the coefficients of (8) become constant per subdomain, and the source term is given by (19). Thus, the equation becomes a linear parabolic PDE, uncoupled from the rest of the equations of the model. Compared to the P2D model, the biggest differences should appear when this source term is far from being spatially homogeneous. For current signals that are piece-wise constant over time, that is especially true the first moments after a change in the current magnitude. When the current signal stays constant, both models reach, eventually, similar steady-state profiles. This is illustrated in Figs. 4 and 5.

3.4 Cell Voltage

In the P2D model, the cell voltage is typically computed as

$$V(t) = \phi_s^+(L, t) - \phi_s^-(0, t) - R_{\text{col}} i(t). \tag{24}$$

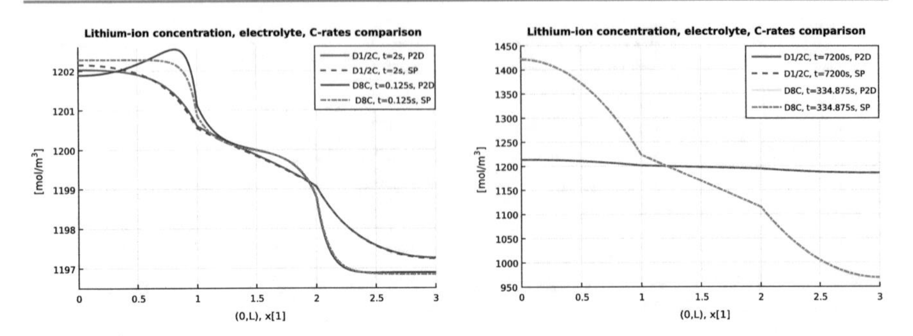

Fig. 4 Discharge under constant 0.5 and 8C-rates. Lithium salt concentration in the electrolyte at the first (left) and the last (right) time step simulated

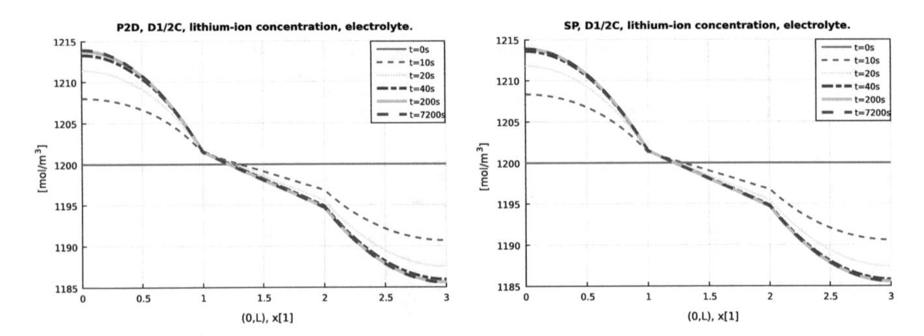

Fig. 5 Discharge under constant 0.5C-rate. Lithium salt distribution across the cell for the P2D (left) and SP (right) models, at different times

For our SPM, we can use (18) and (20) to obtain an explicit expression of voltage as a non-linear function of current and lithium concentration in the different phases. Hence, it is not necessary to solve PDEs (10 – 15). Indeed, from (18),

$$\phi_s^+(L,t) - \phi_s^-(0,t) = \phi_e(L,t) - \phi_e(0,t) - \left(\frac{z_{\text{film}}^+}{L^+ a_s^+} + \frac{z_{\text{film}}^-}{L^- a_s^-}\right)\frac{i(t)}{S} + \eta^+(t) \tag{25}$$

$$-\eta^-(t) + U_{\text{ocp}}^+\left(\frac{c_s^+\left(R_s^+,t\right)}{c_{s,\max}^+}\right) - U_{\text{ocp}}^-\left(\frac{c_s^-\left(R_s^-,t\right)}{c_{s,\max}^-}\right). \tag{26}$$

We recall the assumption that κ^{eff} and κ_D^{eff} are constant per subdomain (A_1). Integrating (10) across Ω, one can deduce the equality

$$\phi_e(L,t) \quad - \quad \phi_e(0,t) = -\frac{2R\theta}{F}\left(t_+^o - 1\right)\left(\ln\left(c_e(L,t)\right) - \ln\left(c_e(0,t)\right)\right) \tag{27}$$

$$-\frac{i(t)}{2S}\left(\frac{L^-}{\kappa^{\text{eff},-}} + \frac{2L^{\text{o}}}{\kappa^{\text{eff,o}}} + \frac{L^+}{\kappa^{\text{eff},+}}\right). \tag{28}$$

Substituting (20), (25 – 26) and (27 – 28) in (24), we finally obtain

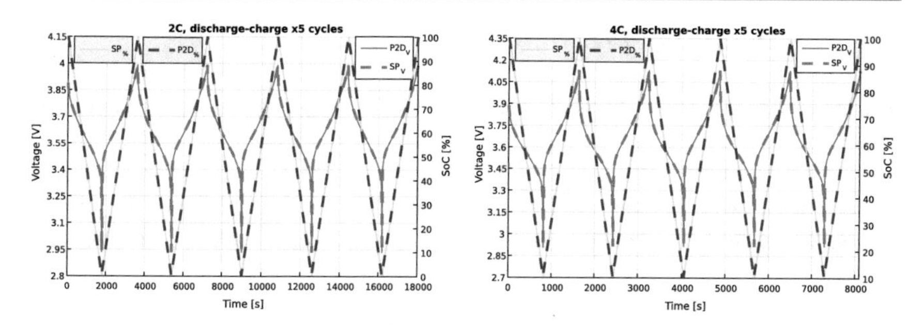

Fig. 6 Cell voltage and state of charge values obtained using the P2D and SP models for 5 discharge-charge cycles, under constant C-rate current signals

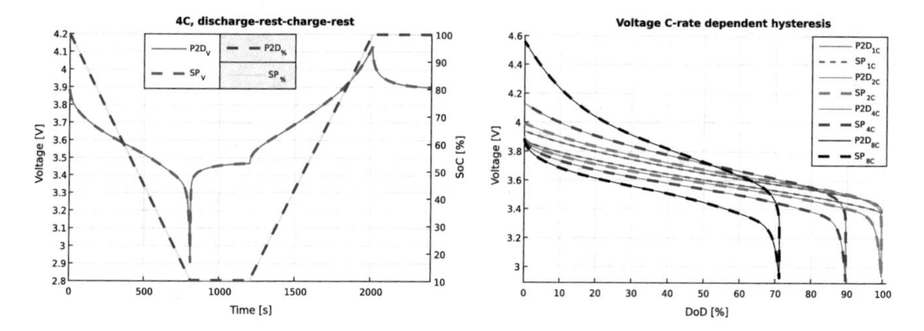

Fig. 7 Cell voltage phenomenology captured by both P2D and SP models. To the left, relaxation effects. To the right, under constant current discharges-charges, voltage C*-rate dependent hysteresis* as a function of the cell deep of discharge, due to the difference between average and surface lithium saturation of solid particles

$$V(t) = \frac{R\theta}{\alpha F}\left(\sinh^{-1}\left(-\frac{i(t)}{2Sa_s^+ i_{0,\mathrm{avg}}^+(t)L^+}\right) - \sinh^{-1}\left(\frac{i(t)}{2Sa_s^- i_{0,\mathrm{avg}}^-(t)L^-}\right)\right) \tag{29}$$

$$-\left(\frac{z_{\mathrm{film}}^+}{L^+a_s^+} + \frac{z_{\mathrm{film}}^-}{L^-a_s^-}\right)\frac{i(t)}{S} - \left(\frac{L^-}{\kappa^{\mathrm{eff},-}} + \frac{2L^o}{\kappa^{\mathrm{eff},o}} + \frac{L^+}{\kappa^{\mathrm{eff},+}}\right)\frac{i(t)}{2S} \tag{30}$$

$$-R_{\mathrm{col}}i(t) - \frac{2R\theta}{F}\left(t_+^o - 1\right)\left(\ln\left(c_e(L,t)\right) - \ln\left(c_e(0,t)\right)\right) \tag{31}$$

$$+U_{\mathrm{ocp}}^+\left(\frac{c_s^+(R_s^+,t)}{c_{s,\max}^+}\right) - U_{\mathrm{ocp}}^-\left(\frac{c_s^-(R_s^-,t)}{c_{s,\max}^-}\right). \tag{32}$$

The SPM expression for the cell voltage accounts for solid particles saturation at their surfaces, the lithium salt gradient in the electrolyte, and materials resistance effects. We find a good agreement between the qualitative behavior of cell voltage computed with the P2D and the SPM, both capturing cell polarization and relaxation effects under high C-rates, Figs. 6 and 7.

3.5 State of Charge of the Cell

The state of charge (SoC) of a cell is a measure of the amount of charge stored in the device.

In physics-based models, it can be expressed in terms of the average lithium saturation of the solid phase, in any of the electrodes. Using the SPM, the average lithium concentration of intercalated lithium in each electrode can be computed as

$$c_{s,\mathrm{avg}}^{\pm}(t) = \frac{3}{R_s^{\pm 3}} \int_0^{R_s^{\pm}} r^2 c_s^{\pm}(r,t)\,\mathrm{d}r. \tag{33}$$

Then, we relatively measure the state of charge of the cell, with respect to the nominal stoichiometry coefficients, as a linear function of $c_{s,\mathrm{avg}}^{\pm}(t)$ by the expression

$$\frac{\mathrm{SoC}(t)}{100} = \frac{\frac{c_{s,\mathrm{avg}}^{-}(t)}{c_{s,\max}^{-}} - x_{0\%}}{x_{100\%} - x_{0\%}} = 1 - \frac{\frac{c_{s,\mathrm{avg}}^{+}(t)}{c_{s,\max}^{+}} - y_{0\%}}{y_{100\%} - y_{0\%}}. \tag{34}$$

The presented models in this work (P2D model and SPM) conserve charge and mass in terms of the current signal. Hence, SoC is equivalently estimated for both, since the SPM just averages the P2D dynamics.

4 Numerical Results

In the previous section, we have discussed some qualitative properties of the SPM using numerical results. We give now the technical and quantitative details of the numerical experiments. We did not aim for an optimal choice of meshes or implementation of the SPM in terms of computational cost and accuracy. The experiments were designed to get an insightful first approach to the SPM features. Every result presented could be potentially improved. Our main goal was to summarize the SPM performance, comparing it with the original P2D model.

The P2D equations present several implementation challenges: different spatial scales domains, non-linear coupled parabolic and elliptic PDEs, non-linear algebraic constraints (Butler–Volmer kinetics), etc. Numerical simulations of this model have been carried out using a *Repsol*[1] & *Itmati*[2] software, temporally ceded for this work [12].

On the other hand, the SPM derived consists of three decoupled, one-dimensional and linear parabolic PDEs. The cell voltage is expressed as a non-linear function of the current and the lithium concentrations value in the different phases. When simulating for a given current signal, one can compute the cell voltage as a post-process, after the numerical resolution of the model. Due to its linearity and simplicity, it presents a

[1] www.repsol.com.

[2] www.itmati.com.

classical variational formulation for parabolic PDEs, using weighted Sobolev functional spaces in the case of spherical PDEs. For its numerical resolution, the finite element method library FEniCS [1] with linear Lagrange elements has been used.

Several charge and discharge protocols with piece-wise constant current signals were considered. For both models, an implicit Euler integrator was used, with a fixed time discretization step, computed as $dt = \frac{1}{n}[s]$, whenever the current applied as input was a nC-rate signal, with $n \in \mathbb{N}$. The parameter's data of the cell models were taken from [20]. Models were run in a laptop computer with processor *Intel(R) Core(TM) i5-6200U CPU @ 2.30 GHz 2.40 GHz* with 8GB RAM memory and 64-bits Windows 10 (OS).

Notice that the SPM does not present a two-dimensional micro-scale domain, as the P2D model does. Therefore, the number of degrees of freedom of the discretized problem is greatly reduced, as shown in Table 2, where we summarize the number of nodes of the meshes involved in the computations. The simplicity of the SPM allows avoiding several not negligible technical difficulties of the P2D model numerical resolution like, for example, the micro-macro scale coupling.

In Table 3, we compare voltage value difference between models and computational times for single discharges under constant C-rate current signals. As expected, the voltage differences norm grows for higher C-rates. But, the relative maximum

Table 2 Mesh data for each model

Subdomain	Dimension	Number of nodes
P2D model		
Ω	1D	300
$\Omega^{\pm}$	1D	100
$(0, R_s^{\pm}) \times \Omega^{\pm}$	2D	10,000
SPM		
Ω	1D	100
$(0, R_s^{\pm})$	1D	100

Table 3 Voltage values and computational times comparison between the P2D and the SP models, under different constant C-rate discharges

	0.5C	1C	2C	4C	8C
Voltage					
Max. error (%)	2.889	2.887	2.885	2.882	2.877
$\|\boldsymbol{V}^h_{\text{P2D}} - \boldsymbol{V}^h_{\text{SPe}}\|_2$	0.0063	0.0123	0.0441	0.0524	0.08598
Resolution time					
SPe	97.9 s	108.0 s	121.8 s	100.9 s	74.0 s
P2D	3246 s	3470 s	3320 s	3279 s	2700 s
Gain	×33	×32	×27	×32	×36

difference is similar for every current signal. Notice that the greatest differences occur at the beginning of the discharge simulations when P2D dynamics are far from being homogeneous at each electrode. The SPM does not capture the reaction flux behavior for those instants, under any C-rate current signal.

5 Conclusions

Assuming homogeneous and known cell temperature, and neglecting degradation mechanisms over-time, a single particle model with electrolyte dynamics has been deduced from a P2D model. We have stated the necessary simplifying assumptions for the model derivation and exemplified their consequences with numerical experiments.

The SPM presents a linear formulation of uncoupled one-dimensional parabolic PDEs and, hence, its implementation is straightforward. Compared to the P2D model, the number of degrees of freedom of the discretized problem can be significantly reduced, while obtaining equivalent estimations of the cell state of charge, and a good agreement for the voltage estimation under high C-rate constant current protocols. With a maximum of, approximately, 3% difference in voltage estimation, the computational cost was reduced up to 27–36 times. All this justifies the SPM potential as a physics-based model of reduced complexity for real-time applications.

Acknowledgements Under the academic supervision of Dr. Jerónimo Rodríguez García, this work was carried out as a Master's thesis for the *Máster en matemática industrial* program, offered by *Universidade de Santiago de Compostela* (www.m2i.es). The project was proposed by *Itmati*, with *Repsol* collaboration, who temporally ceded software to obtain some of the numerical results presented.

References

1. Alnæs, M., Blechta, J., Hake, J., Johansson, A., Kehlet, B., Logg, A., Richardson, C., Ring, J., Rognes, M.E., Wells, G.N.: The fenics project version 1.5. Arch. Numer. Softw. **3**(100) (2015)
2. Baba, N., Yoshida, H., Nagaoka, M., Okuda, C., Kawauchi, S.: Numerical simulation of thermal behavior of lithium-ion secondary batteries using the enhanced single particle model. J. Power Sour. **252**, 214–228 (2014)
3. Balakrishnan, P., Ramesh, R., Kumar, T.P.: Safety mechanisms in lithium-ion batteries. J. Power Sour. **155**(2), 401–414 (2006)
4. Biensan, P., Simon, B., Peres, J., De Guibert, A., Broussely, M., Bodet, J., Perton, F.: On safety of lithium-ion cells. J. Power Sour. **81**, 906–912 (1999)
5. Cai, L., White, R.E.: Reduction of model order based on proper orthogonal decomposition for lithium-ion battery simulations. J. Electrochem. Soc. **156**(3), A154–A161 (2009)
6. Chu, Z., Plett, G.L., Trimboli, M.S., Ouyang, M.: A control-oriented electrochemical model for lithium-ion battery, part i: Lumped-parameter reduced-order model with constant phase element. J. Energy Storage **25**, 100828 (2019)
7. Di Domenico, D., Stefanopoulou, A., Fiengo, G.: Lithium-ion battery state of charge and critical surface charge estimation using an electrochemical model-based extended kalman filter. J. Dyn. Syst. Meas. Control **132**(6), 061302 (2010)

8. Diouf, B., Pode, R.: Potential of lithium-ion batteries in renewable energy. Renew. Energy **76**, 375–380 (2015)
9. Doyle, M., Fuller, T.F., Newman, J.: Modeling of galvanostatic charge and discharge of the lithium/polymer/insertion cell. J. Electrochem. Soc. **140**(6), 1526–1533 (1993)
10. Fan, G., Li, X., Canova, M.: A reduced-order electrochemical model of li-ion batteries for control and estimation applications. IEEE Trans. Veh. Technol. **67**(1), 76–91 (2017)
11. Forman, J.C., Bashash, S., Stein, J.L., Fathy, H.K.: Reduction of an electrochemistry-based li-ion battery model via quasi-linearization and pade approximation. J. Electrochem. Soc. **158**(2), A93–A101 (2011)
12. Giráldez, D.A., Cao-Rial, M.T., Muiños, P.F., Rodríguez, J.: Numerical simulation of a li-ion cell using a thermoelectrochemical model including degradation. In: European Consortium for Mathematics in Industry, pp. 535–543. Springer (2016)
13. Han, X., Ouyang, M., Lu, L., Li, J.: Simplification of physics-based electrochemical model for lithium ion battery on electric vehicle. part i: Diffusion simplification and single particle model. J. Power Sour. **278**, 802–813 (2015)
14. Kennedy, B., Patterson, D., Camilleri, S.: Use of lithium-ion batteries in electric vehicles. J. Power Sour. **90**(2), 156–162 (2000)
15. Li, J., Adewuyi, K., Lotfi, N., Landers, R.G., Park, J.: A single particle model with chemical/mechanical degradation physics for lithium ion battery state of health (soh) estimation. Appl. energy **212**, 1178–1190 (2018)
16. Li, J., Lotfi, N., Landers, R.G., Park, J.: A single particle model for lithium-ion batteries with electrolyte and stress-enhanced diffusion physics. J. Electrochem. Soc. **164**(4), A874–A883 (2017)
17. Lin, X., Kim, Y., Mohan, S., Siegel, J.B., Stefanopoulou, A.G.: Modeling and estimation for advanced battery management. Ann. Rev. Control Robot. Auton. Syst. **2**, 393–426 (2019)
18. Lu, L., Han, X., Li, J., Hua, J., Ouyang, M.: A review on the key issues for lithium-ion battery management in electric vehicles. J. Power Sour. **226**, 272–288 (2013)
19. Luo, W., Lyu, C., Wang, L., Zhang, L.: A new extension of physics-based single particle model for higher charge-discharge rates. J. Power Sour. **241**, 295–310 (2013)
20. Mazumder, S., Lu, J.: Faster-than-real-time simulation of lithium ion batteries with full spatial and temporal resolution. Int. J. Electrochem. **2013** (2013)
21. Moura, S.J., Argomedo, F.B., Klein, R., Mirtabatabaei, A., Krstic, M.: Battery state estimation for a single particle model with electrolyte dynamics. IEEE Trans. Control Systems Technol. **25**(2), 453–468 (2016)
22. Ning, G., Popov, B.N.: Cycle life modeling of lithium-ion batteries. J. Electrochem. Soc. **151**(10), A1584–A1591 (2004)
23. Perez, H., Dey, S., Hu, X., Moura, S.: Optimal charging of li-ion batteries via a single particle model with electrolyte and thermal dynamics. J. Electrochem. Soc. **164**(7), A1679–A1687 (2017)
24. Plett, G.L.: Battery management systems. In: Battery Modeling, vol. 1. Artech House (2015)
25. Plett, G.L.: Battery management systems. In: Equivalent-Circuit Methods, vol. 2. Artech House (2015)
26. Rahimian, S.K., Rayman, S., White, R.E.: Extension of physics-based single particle model for higher charge-discharge rates. J. Power Sour. **224**, 180–194 (2013)
27. Reniers, J.M., Mulder, G., Howey, D.A.: Review and performance comparison of mechanical-chemical degradation models for lithium-ion batteries. J. Electrochem. Soc. **166**(14), A3189–A3200 (2019)
28. Santhanagopalan, S., Guo, Q., Ramadass, P., White, R.E.: Review of models for predicting the cycling performance of lithium ion batteries. J. Power Sour. **156**(2), 620–628 (2006)
29. Santhanagopalan, S., White, R.E.: Online estimation of the state of charge of a lithium ion cell. J. Power Sour. **161**(2), 1346–1355 (2006)
30. Sharma, A.K., Basu, S., Hariharan, K.S., Adiga, S.P., Kolake, S.M., Song, T., Sung, Y.: A closed form reduced order electrochemical model for lithium-ion cells. J. Electrochem. Soc. **166**(6), A1197–A1210 (2019)

31. Subramanian, V.R., Ritter, J.A., White, R.E.: Approximate solutions for galvanostatic discharge of spherical particles i. constant diffusion coefficient. J. Electrochem. Soc. **148**(11), E444–E449 (2001)

Fracture Propagation Using a Phase Field Approach

David Casasnovas and Ángel Rivero

Abstract

Phase field models have received a lot of attention during the last 20 years and they have reached maturity, being in the last years ubiquitous, finding nice examples in a range of applications in physical sciences and engineering, from the classical spinodal decomposition in multiphase flows to qualitative studies of motility in metastatic tumor cells. In this paper, we give a brief introduction to the theory of the method and present a review of some striking applications in order to show the huge potential and versatility of the technique. In this work, we are interested in very specific problems related to applications in energy storage and fracture dynamics. This research began by using a simple model to study the propagation of fractures in elastic homogeneous materials and eventually evolved into a coupled model that includes fracture propagation and flow in elastic-porous media.

1 Introduction

The phase field model is a macroscale mathematical model for solving problems with interfaces that define two or more equilibrium states of the system. It can be seen as a macroscopic version of the Density Functional Theory (DFT), which is one of the most successful models in solid-state physics, molecular electronics, or quantum chemistry. In short, we search for an extrema of a functional (e.g., free

D. Casasnovas · Á. Rivero (✉)
TechLab Repsol, Advanced Mathematics Group, Madrid, Spain
e-mail: angel.rivero@repsol.com

D. Casasnovas
e-mail: d.casasnovas@repsol.com

P. Quintela Estévez et al. (eds.), *Advances on Links Between Mathematics and Industry*, SxI - Springer for Innovation / SxI - Springer per l'Innovazione 15,
https://doi.org/10.1007/978-3-030-59223-3_7

energy) which depends on one or several "order parameters" (also referred to as "phase fields") to be determined. The functional should contain all the relevant contributions to the energy that can change during the evolution of the system along the energy landscape, which very often shows several local minima. How the system evolves along this landscape would be the result of the model we would like to derive but as it is stated, the solution is the order-parameter fields in the minimum energy configuration. The evolution equation is usually derived from the variational formulation resulting in Cahn–Hilliard or Allen–Cahn formulations, or in variations of these.

In this work, we present a variety of examples and results of our own simulations and experience using phase field. All of them are own results except where it is otherwise indicated and they use a diffuse-interphase model together with the phase field model. In classical models, the interface between two fluids is treated as infinitely thin (sharp) and is assumed to have physical properties such as surface tension. For instance, the Stefan problem involves solving a free boundary that couples two fields to be determined. On the contrary, in diffuse-interface theories, this sharp interface[1] is replaced with continuous variations of the order parameters and is usually seen as a regularization of the sharp-interface problem.

The phase field method might represent real physics in some cases, while in others the method might be better viewed just as a suitable computational technique. In any case, phase field models should be understood as a coarse-grain model because at fine resolution, neither phase field models nor sharp-interface models perfectly represent the physical systems, and molecular dynamics simulations should be used instead.

Despite their limitations at the microscopic level, in the continuous regime phase field models have clear advantages. Since the introduction of the phase field theory in the 50s [6] and the generalization in DFT for microscopic systems in the 60s [13] the number of applications has been increasing and the number of articles and excellent reviews in the last 20 years on this topic is enormous [3, 22].

2 The Phase Field Model

First, one should be familiar with the concept of an "order parameter" or a "phase field". It is not easy to give a general definition that includes all possible cases. Sometimes a single scalar order parameter suffices. In other cases, we need multiple parameters. Intuitively, a single phase field variable is a way to label two different regions of the system with different intrinsic properties at macroscopic level.

The simplest example is the heterogeneous mixture of two immiscible fluids, e.g., oil and water in a pan [14]. When the oil–water mixture is stirred and isolated, we observe the formation of small oil droplets that immediately coalesce to form

[1] A diffuse-interface model with a finite width that tends to zero can be considered as a sharp interface but it must not be confused with the so-called thin interface limit.

larger droplets and gradually segregate (unmix) from water. After a long enough time lapse, a unique big oil drop floats in water. In multiphase fluid systems, one can consider the phase field variables to be the volume fraction of the phases. Generally, to describe the system one needs the same number of parameters as the number of phases minus one. This is because the sum of all the phases is a fixed quantity (i.e., the sum of all the volume fractions is unity).

In a polycrystalline solid where several crystalline domains share the volume, the lattice of each crystal has different properties such as atom spacing, orientations or densities. Here each crystalline domain corresponds to a different value of the phase field and we would need again as many order parameters as different crystalline systems we have minus one.

Going back to a multiphase system such as humid air and liquid water, one considers a box with this system, initially in equilibrium, in a homogeneous mixture of air and vapor everywhere. This mixed state of molecules of nitrogen, oxygen, and water corresponds to a unique fluid phase, that can be labeled, e.g., $\phi = 0$, since the length scale at which the molecular differences arise is far below the level of description of the model. So, from a practical point of view the gas mixture is a single phase. In addition consider that at some walls, if the system is undercooled (i.e., ready for a metastable phase transition) and the wavelength of the fluctuations of density or pressure is large enough, the vapor in the air condensates through heterogeneous nucleation forming small drops on the walls. The condensed phase is composed mostly by water molecules and it forms an own phase, let's label it $\phi = 1$.

Therefore, if one can obtain consistent relationships of the different phenomena such as multicomponent thermodynamics of heterogeneous systems in the above examples, in terms of a scalar field (hereafter, the phase field) that in most of the domain it takes the value either 0 or 1 and intermediate values in the transitions or interfaces, then we are in a position to get a physically realistic description of the evolution of the system. We'll return to this shortly in greater detail.

Sometimes, as in the second example, a single parameter is not enough because the system has two phases. In this work, we consider cases with a single scalar parameter. However, we will explore other complexities with a non-conventional understanding of "phase" (Fig. 1).

To better understand the method, we explain it with the simplest case of the classical model for an isothermal spinodal decomposition[2]. For the sake of simplicity, consider that for an initial condition $t = 0$, the fluids form a unique phase and at $t > 0$ the condition changes abruptly (e.g., the stirring ceases or the pressure drops) to another where they become immiscible. The value labels assigned to the phases are completely arbitrary but usually either 0 and 1, or −1 and 1 are chosen by convention. But once the choice is made, one must bear in mind that it affects the

[2]The initial fluid might be composed by different components mutually soluble that under such abrupt change suddenly losses the solubility condition and the components separate into different phases. Salted water or brine is a simple binary example. A hydrocarbon mixture is a more complex multicomponent example.

Fig. 1 Three applications of the phase field: separation of mixed phases, anisotropic solidification, and wettability in solid walls

specific form of the energy potential and subsequent derived expressions. It is also important to note that the phases are reorganized but the amount of each one is constant along the process, i.e., there is no phase change in this case.

Hereafter, the system at any time is described by an order parameter $\phi(x)$ such that $\phi = 0$ for the region occupied by one fluid or phase (e.g., oil) and $\phi = 1$ elsewhere (e.g., water). Initially $\phi = \phi_o$ everywhere (e.g., mixture), being between 0 and 1. In some cases an amount of fluctuations (amplitude and size) is needed to trigger the phase separation as the model reproduces phase separation by an instability process. A simple linear stability analysis shows that depending on the value of ϕ_o one may need a different amount of fluctuation to surpass the energy barrier between the two wells [24]. The general dispersion relation is $\omega = k^2(1 - a^2k^2)$, where k is the wavenumber and a relates to the mean level of fluctuations $\phi_{o.}$ The exact expression for a depends on the form of the dimensionless equation but always is quadratic in ϕ_o and is important for the metastability of all these systems. The key is that a range of unstable wavenumber (i.e., with amplification ω greater than zero) exists with maximum instability at $k \sim a/\sqrt{2}$.

Going back to the oil–water separation, if one considers a fixed volume box, the initial mixture will evolve along a trajectory in the free energy landscape toward a global minimum. In this simple example, the term *landscape* might be pretentious, as the system just moves on a plain curve (Fig. 2) but the term becomes meaningful as we add ingredients introduced in the next sections. The total free energy must include, in addition to the energy contained into the different patches of water and oil, the non-negligible interfacial energy (i.e. surface tension) through terms containing $grad(\phi)$ that accounts for the energy in the interface, as it will be shown shortly. The spinodal decomposition (i.e., the demixing process) occurs for any quench that leaves the system beneath the spinodal line and immediately after the system becomes unstable.

The free energy, for any process, will be lowered by creating two domains whose distribution is unknown and part of the solution but in order to describe quantitatively the domains and the interfaces between them, one must specify not just the volume free energy $f(\phi)$, but the free energy functional, $F[\phi]$

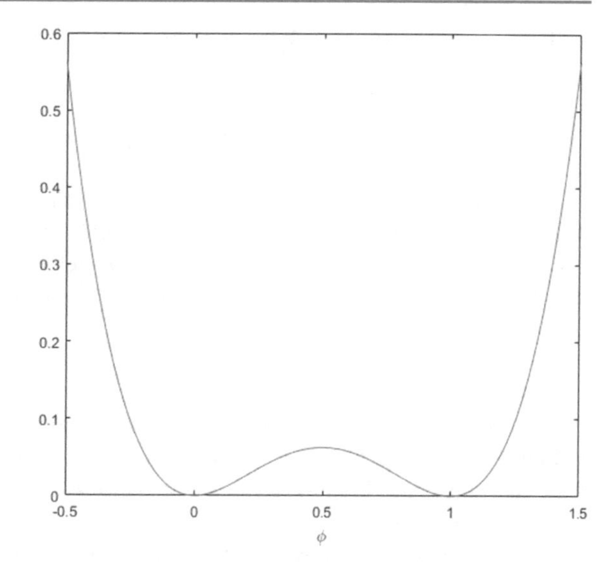

Fig. 2 Typical double-well potential used in most of the phase field models for $f(\phi)$, the bulk free energy density

$$\underset{\phi}{argmin}\, F[\phi(x), \nabla\phi(x)]$$

$$F[\phi(x), \nabla\phi(x)] = \int \left(f(\phi) + \tfrac{\varepsilon^2}{2}|\nabla\phi|^2\right) dV \qquad (1)$$

The integral (1) extends on the whole domain. Usually the bulk free energy density f is a double-well potential corresponding to the two equilibrium states that compete in the system. The exact form of the potential is not very important for qualitative applications or near critical conditions but in specific situations it links directly to the real thermodynamic equation of state (let be,Van der Waals, Peng Robinson, or any other in real systems) and it requires some care in the connection. Also, issues related to the double-well function in the spontaneous drop shrinkage phenomenon has been reported in the literature and alternative energy functions have been proposed [11].

The ε parameter in the functional is related to the numerical interface thickness which is a parameter of the model and is, in most of the situations, much larger than the physical thickness, in the order of nanometers. In the cases where this dissimilarity is important, then both thicknesses should be of the same order and we are forced to use an extremely fine mesh which may be prohibitively expensive.

Once the free energy functional is defined we need to find the driving force so that the system evolves. Usually this part is the trickiest step in the procedure, and it is very dependent of the type of problem we are studying. Roughly speaking we must find some relationship between conservation laws and the fluxes of the different magnitudes expressed in terms of variables (in fact, variational derivatives) derived from the free energy such as chemical potentials.

In the oil–water example it is necessary to begin with the generalized form of the Fick's law

$$\overrightarrow{J_\alpha} = -M_\alpha \nabla \mu_\alpha \tag{2}$$

J_α is the flux of the phase α, M_α is its mobility and μ_α is its chemical potential (α is either water or oil). The mobility is related with the diffusion of that phase and we can consider it as a constant for the sake of simplicity. Now we use the definition of the chemical potential field in terms of the functional[3] derivative of the free energy

$$\mu_\alpha = \frac{\delta F}{\delta \phi_\alpha} \tag{3}$$

One can think of the order parameter ϕ_α as the molar fraction of the α-phase and the sum for all the phases (only 2 in this case) is unity so we only need one evolution equation (because $\phi_\beta = 1 - \phi_\alpha$). Hereafter we drop the sub-index and use just ϕ instead.

The variational derivatives are found by applying the Euler–Lagrange equation to the energy functional plus additional constraints that we can impose to the system such as conserved magnitudes [1]. Finally, the dynamics of the order parameter is governed by the mass conservation law

$$\frac{\partial \phi}{\partial t} = -\nabla \cdot \vec{J} \tag{4}$$

Substitution of Eqs. (2) and (3) produces finally the Cahn–Hilliard or the phase field equation for an immiscible two phase system

$$\frac{\partial \phi}{\partial t} = \nabla \cdot M \nabla \frac{\delta F}{\delta \phi} \tag{5}$$

In summary, the procedure consists in defining a functional for the energy of the system including all the contributions that we think can change along its evolution on the energy landscape. The extrema of the functional are the equilibrium states at which the system will evolve and variational derivatives of the functional are usually related to some thermodynamical variables whose gradients are fluxes of quantities that force the system to its equilibria.

2.1 Extended Models

The aforementioned basic framework, namely, a heterogeneous system with two equilibrium states, is a stepping stone to introduce more complex systems. The method has seen an explosion of applications in the previous years and it would be naïve and pretentious just to want to list them. So we limit ourselves to give several representative examples in our domain expertise, explaining the most remarkable

[3]Functional or variational derivative in the sense used in calculus of variations [9].

features of those models in the different applications as well as citing some of the latest and most impressive work to our knowledge. All the examples shown in the next sections have been made using our own models and solving them with in-house codes except the model of metastasis in tumor cells.

2.1.1 Three Phases Systems

The first straightforward extension is to add a third immiscible phase (e.g., water–oil–air) by using a second parameter [4, 5]. The free energy landscape now shows three different minima and it can be obtained either from thermodynamical considerations or from heuristics. In both cases, the free energy density potential should verify certain compatibility conditions so that the results be physically relevant.

In particular, the model must reduce to the biphasic Cahn-Hilliard model (described in previous section) when only two phases are present in the mixture. Then it is said that the model is *algebraically* consistent with binary systems. Furthermore, the model should be also *dynamically* consistent with binary systems that ensures that non-physical emergence of one spurious phase inside the interface between the other two will not happen.

With three phases arise contact lines—geometrical places where all three phases coexist. The relative affinity between different fluids will determine the equilibrium angle and physically is given by the energy contained in the interfaces, i.e., the interfacial tension. In the model usually *spreading coefficients* Σ_α for each phase are used instead and are defined in terms of the surface tension $\sigma_{\beta\gamma}$ of the other two phases. In Fig. 3, three cases are shown with different spreading coefficients that result in contact angles between phases. Similar to the two-phase case, it is possible to study the dynamics of three phases that are initially mixed and how they separate into three distinct phases depending on their interfacial tensions (Fig. 4).

2.1.2 Confinement and Convection

We now describe how to introduce walls in the formulation. Two fluid phases in the presence of a solid wall show different affinity to wet it and its effect is measured through the contact angle. In this way, a neutral wetting (90º contact angle) indicates similar trend in both fluids. The contact angle is a macroscopic reflection of the averaged effect of the weak interactions between molecules of the fluids and the superficial molecules of the solid. Now it is clear that the energy f_S of those transient fluid–solid molecular bonds must be introduced in some way into the variational formulation.

The simplest approximation is to assume that f_S is linear in the local composition ϕ_S, i.e. ϕ along the wall

$$f_S = H\phi_S \tag{6}$$

How fast f_S changes as we move across the contact line is directly related to the contact angle and an expression can be inferred if we minimize jointly the free energy functional (1) plus the wall energy contribution (6)

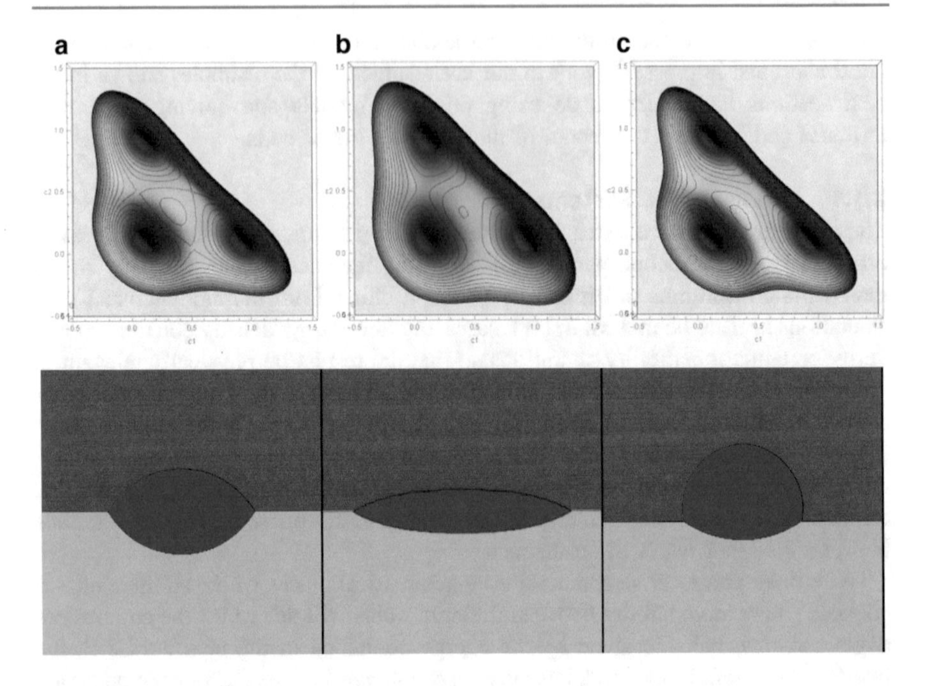

Fig. 3 Free energy density landscapes (top) and their correspondent equilibrium state (bottom) **a** three symmetrical phases, $(\Sigma_1, \Sigma_2, \Sigma_3) = (4, 4, 4)$, **b** phase 2 and 3 symmetrical, (4, 2.4, 2.4), **c** all three asymmetrical (4, 3.2, 5.6)

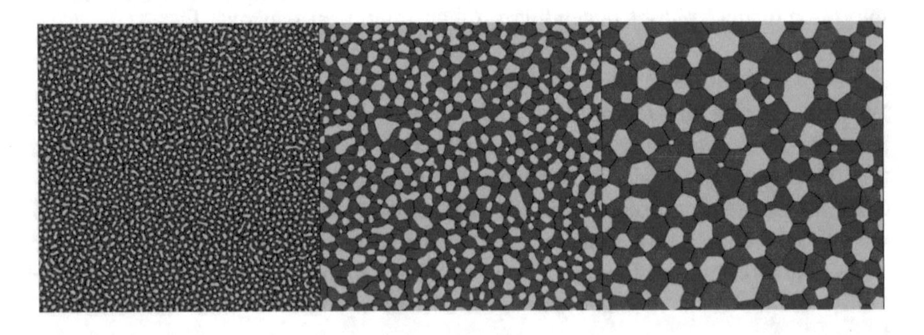

Fig. 4 Snapshots of the spinodal decomposition for three symmetrical phases (case a, in Fig. 3) from an initial uniformly random distribution of phases

$$|\nabla\phi \cdot \vec{n}| = \frac{2}{\varepsilon^2}\frac{df_S}{d\phi_S} = \frac{2H}{\varepsilon^2} \tag{7}$$

A complete-wetting to partial-wetting transition can be obtained by tuning the specific energy H. Further details are given in [10, 16].

Fig. 5 Snapshots of the separation process in a channel. The walls are up and down, low surface tension (5e-4 N/m) and the wall is highly wetted (CA = 10º) by the blue fluid

Figure 5 illustrates the dynamics of the spinodal decomposition between two highly "hydrophilic" walls (the blue fluid represents "water"). It is also assumed that the channel width and drops are small enough so that we can omit gravitational effects. Figure 6 shows the same scenario but the fluids have different properties.

Fig. 6 (Same as Fig. 5) High surface tension (5e−2 N/m) and the wall is mid-wetted (CA = 50º) by the blue fluid

We now introduce convection which converts the method in a truly powerful and versatile tool in many applications. The velocity field can be given as in some microfluidic or small-scale applications where a parabolic profile is used and a one-way coupling is assumed. In this case we are interested in a complete coupling with the velocity field that is computed solving simultaneously the Navier–Stokes equations. In that case, the phase field is simply advected by the flow while the effect of the phase dynamics on the flow is introduced through capillarity. The capillarity forces usually are computed using the gradient of chemical potential (which enables us to study the Marangoni effect) and the phase field.

Using such a coupled model we can compute multiphase flows in complex geometries. In reservoir engineering, for example, we extract the microstructure at pore level from real three-dimensional samples of rocks using micro Computerized Tomography and simulate the displacement of fluids (Fig. 7). Another important ingredient in macroscopic flows is buoyancy in cases with macroscopic mixing of fluids of different density coming from Rayleigh-Taylor instabilities (Fig. 8).

2.1.3 Non-Isothermal Models and Anisotropy

The application of phase field models to study solidification processes in metallurgy dates to the 60s and the work of Cahn and co-workers. The problems involving phase change in multicomponent fluids with moving interfaces imply that some

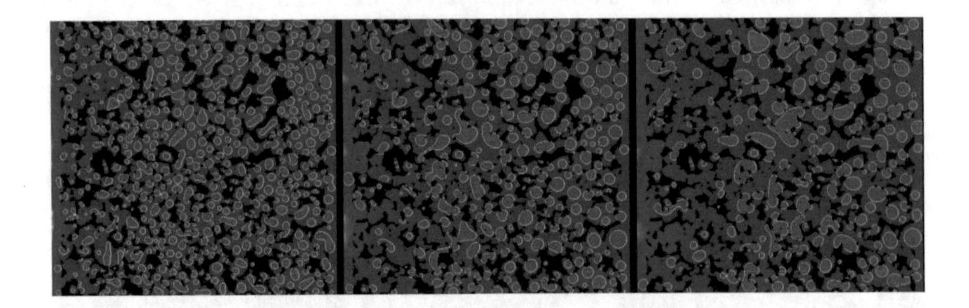

Fig. 7 Snapshots of the dynamics of multiphase displacement in a two-dimensional synthetic rock. Initially the rock is filled by a macroscopic emulsion and blue fluid comes in from left. The rock here is blue-wet (CA = 50°), velocities are 1 mm/s and viscosity ratio is 1

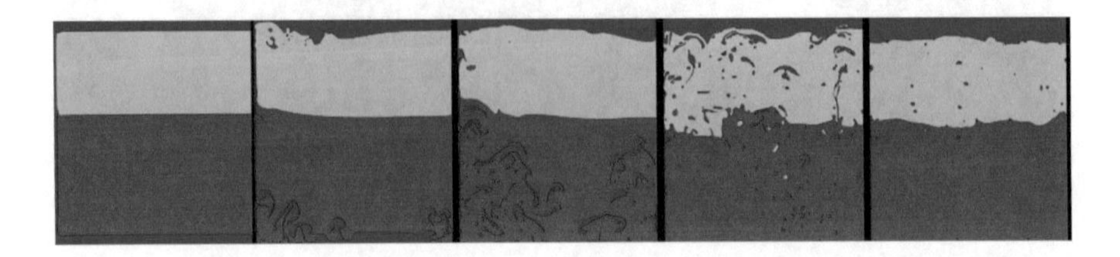

Fig. 8 Snapshots of the dynamics of mixing in a three-phase buoyant flow with hypothetical fluids. Denser fluid is red, ligther one is blue and density ratio is 10

components migrate between phases according to the local thermodynamics. Generally, in order to include the interface kinetics in the phase field model requires that we add the migrating component's field and its evolution.

The construction of the free energy functional in terms of phase and components is at the core of the method and the connection with a specific thermodynamic system is evident. There are several ways to construct the free energy function from thermodynamic considerations [3, 23]. In some cases, only one component (or several ones but with similar kinetics) is involved in the phase change, e.g., precipitation of salts, pure material melt, etc. Therefore, we can dismiss the tracking of the different components and simplify the formulation using only one order parameter. In the following, the discussion focuses in this simpler case.

In addition, the model considered so far is isothermal (the free energy density landscape is for a given temperature) and conservative for the phase. There are two approaches to include temperature and phase changes in our model. The first and more rigorous one uses the entropy functional as a function of energy and phase field [17]. The criteria to ensure a positive production of entropy during the process naturally leads us to the equations for energy diffusion and phase evolution.

A simpler approach [15] is to use an isothermal formulation with the free energy expressed in terms of phase field and temperature. In addition, we must convert the conservative Cahn–Hilliard equation for the phase into a non-conservative version —the Allen–Cahn equation- that allows formation of new phases.

The conservative Eq. (5) for constant mobility M becomes

$$\frac{\partial \phi}{\partial t} = M\nabla^2\left(\frac{\delta F}{\delta \phi}\right) \tag{8}$$

but, as mentioned earlier, this conservative form is not useful for this problem and the Allen–Cahn equation is used instead. The simplest evolution equation that ensure a decrease in the free energy is given by

$$\frac{\partial \phi}{\partial t} = M\left(\frac{\delta F}{\delta \phi}\right) \tag{9}$$

Please note that M has changed units. As mentioned before, the chemical potential $\frac{\delta F}{\delta \phi}$ now depends on the temperature, and the enthalpy equation has to be solved simultaneously

$$\frac{\partial h}{\partial t} = \nabla \cdot k\nabla T + L\frac{\partial \phi}{\partial t} \tag{10}$$

where the enthalpy density h includes contributions from both phases, k is the thermal conductivity and the last term corresponds to a source of heat due to the latent heat (L) released at the moving interfaces where the phase ϕ changes with time.

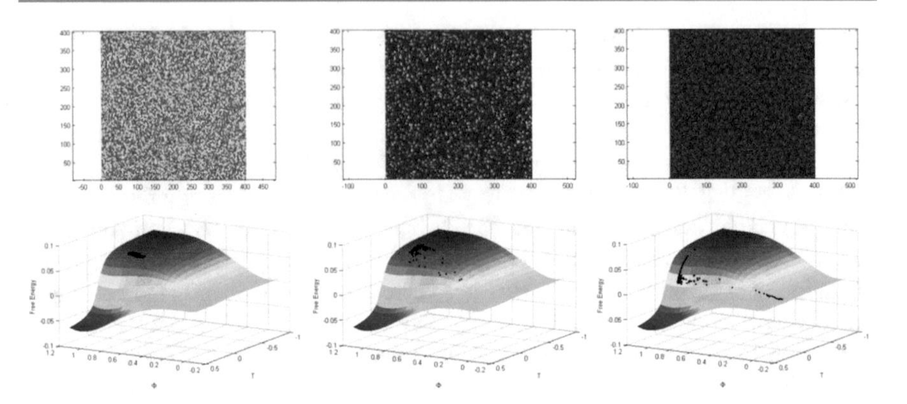

Fig. 9 Growth of solid nuclei in an undercooled melt in adiabatic condition, phase field evolution (top row) and landscape of bulk free energy on which the system evolves (bottom row). Initial state contains small amplitude perturbations at To < Tc ((Tc − To)/Tc = 0.01). As the nuclei grow they heat and the growth ends at Tc

In Fig. 9, we see the nucleation and growth process when a solid phase forms due to local undercooling at temperatures T below a critical T_c. The system evolves along the free energy landscape as the nuclei grow.

Crystal growth is an interesting problem that also can be formulated with phase field models. Actually, it is only necessary to add anisotropy to the previous liquid–solid phase change framework. Of course, it is possible to add more subtle effects, e.g., gravitational or Marangoni effects, but it is out of the scope of this review. Roughly speaking, crystal and dendrites arise when the solidification process occurs at a different rate in different directions. The energy barrier to phase change is higher in some directions than the others. This is the motivation to introduce anisotropy in the model and the simplest way [15] is to assume that ε in Eq. 1 depends on the direction of the outer normal vector of the interface

$$\varepsilon(\theta) = \varepsilon_o(1 + a\cos(j\theta)) \tag{11}$$

Here a measures the strength of anisotropy and j is the angular mode (θ is zero in the horizontal direction). In Fig. 10 we show some examples of crystal growth with two different values of a.

2.1.4 Other Examples

Other interesting applications that can be simulated using phase field models and include electrical fields and electrochemical effects are found in electrodeposition, battery electrodes, electrolysis, and corrosion.

Bazant and co-workers [2, 7, 8] have developed the mathematical framework to guarantee the consistency of all the physical mechanisms involved. On one hand, Faradaic reaction kinetics must be included in the formulation by matching the velocity of the interface at a given voltage, usually by using some

Fig. 10 Phase (top) and temperature field (bottom) for crystals growth from a small nucleus at the middle of the bottom boundary. At left $a = 0.01$ and at right $a = 0.05$

current-overpotential equation (Butler–Volmer eq.). In addition, we need to add the electrical contribution to the free energy. The resultant evolution equations naturally contain the usual diffusive fluxes plus the electromigration flux. Non-physical solute trapping can be an issue and care must be taken by adding some anti-trapping fluxes to the phase field equation.

Recent work [20, 21] by Moure and Gomez reproduces the cellular migration as in amoeboid manner in complex geometries mimicking the motility acquired by tumor cells in metastasis processes. The model is composed by three submodels and includes an impressive number of coupled mechanisms, all of them in the phase field framework. The whole model accounts for the cell motion, the cytosol biochemomechanics (internal extension and contraction of the myosin filaments mediated by actin) and the activator dynamics.

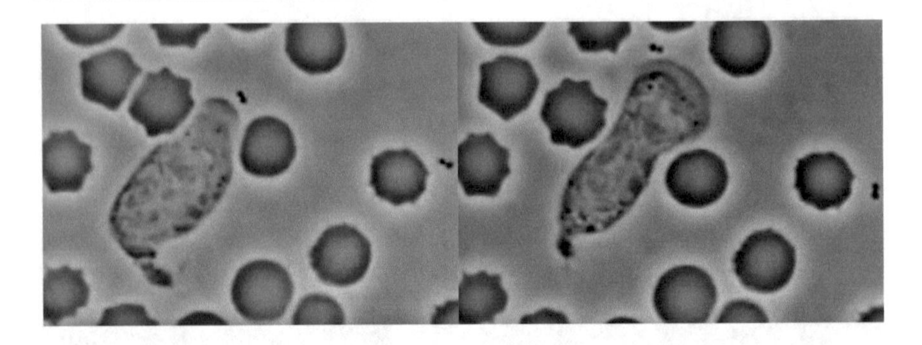

Fig. 11 Blood cell chasing bacteria (small black dots) between red blood cells (from 16 mm movie by D. Rogers, 1950)

The cell detects obstacles around by activating a chemical in that region of the membrane closer to the obstacle. The gradient of the activator triggers the internal kinetics of the myosin-actin system and gives rise to the growth of pseudopods and ease the migration mimicking the real behavior of living cells (Fig. 11).

3 Fracture Dynamics

The formation and propagation of fractures in solids is an interesting and ubiquitous problem. Traditionally they are considered as a singularity, given that at the relevant scale the width of the fracture is much smaller than the dimension of the solid in which it propagates.

At a given time the solid can be divided into a damaged region (fractures) and into a "healthy" one. This insight links directly with the core of the phase field models where several equilibrium states of a free energy landscape compete between them.

In the following, we develop the guidelines to apply the diffuse-interface approximation and the phase field model to solve the problem. The evolution of the fracture is seen as a path to the minimum energy configuration and the resolution of the displacement field with a given fracture is complemented by the equation of evolution of the phase field that determines the damaged regions (fractures).

An important assumption in the derivation of the model is that the variation of the boundary conditions (applied stresses, pressure or massflow) is slow compared to that of fracture propagation and of restoring the equilibrium stress field. This situation is common since the stresses propagate at the speed of sound and the fracture travels to a fraction of it, typically of the order of 1000 m/s and 300 m/s, respectively. Then the equations are reduced to the calculation of the stationary solution.

In particular, we aim to solve the problem of hydraulic fracturing, namely, the propagation of a fracture in a porous granular and confined media, like a rock, driven by fluid injection in a certain region. The rock is permeable and both the rock and the fluid are compressible. The main physical assumptions are

- Crack irreversibility that avoids the healing of the fracture.
- Elastic regime in solids as in brittle materials like rocks.
- Compression energy does not contribute to crack growth. The fractures are created and propagated by shear or traction [12].
- Permeability has two contributions: the bulk and the one related to the crack. The latter depends on the crack width.
- The value $\phi = 1$ corresponds to the crack and $\phi = 0$ to the healthy rock.

The mathematical formulation of the hydraulic fracturing of fluid-saturated porous media is composed by several mechanisms: elastic rock mechanics, compressible fluid flow in porous media, and crack propagation model (phase field).

$$\mathrm{div}(\boldsymbol{\sigma}) = 0,\ \text{where}\ \sigma = \boldsymbol{\sigma_{\mathrm{eff}}} - bp\boldsymbol{I} \tag{12}$$

$$\frac{1}{\mathrm{M}}\dot{p} + b\nabla\cdot(\dot{\boldsymbol{u}}) - \nabla\cdot(\boldsymbol{K}\nabla p) = \mathrm{Q} \tag{13}$$

$$K = k_{rock}I + \phi^{\epsilon}K_{crack} \tag{14}$$

The elasticity is modeled using the Biot's equations for linear poroelasticity, where $\boldsymbol{\sigma}$ stress decomposes into the Biot's effective stress $\boldsymbol{\sigma_{\mathrm{eff}}}$ and the stress due to pore pressure (12). The Biot's effective stress coefficient, b, describes the change of the bulk volume due to a pore pressure change while the stress remains constant.

The model only captures the time scales associated to the pressure and flow. The fluid flow in porous media is modeled by the single-phase Darcy's law, the continuity equation and compressibility of the media. The pressure field (13) is computed for a given mass flow of fluid, modeled as a source term Q. Here M is the Biot's modulus (accounts for the compressibility of the fluid) which is defined as the fluid volume change per unit control volume and per unit pressure change, keeping the control volume constant and $\boldsymbol{K}$ is the combined permeability (bulk plus crack).

The phase field is computed from the energy functional accounting for the bulk elastic contribution W_u and the surface energy in the crack. By subjecting the dissipation be positive we get the condition

$$2\varphi_{\mathrm{c}}\left[\phi - l_d^2\Delta\phi\right] = 2(1-\phi)H(x,t) \tag{15}$$

Further details can be found in [18, 19]. Equation (15) represents the transition between two states; the unbroken and the fully broken case. In this case, the field depends on a source term H that represents the energy history function over the critical fracture energy density of the porous matrix φ_c that is related to a critical fracture stress

$$H(x,t) = \max_{t \in [0,T]} \{\varphi_{eff}(x,t) - \varphi_c\}$$

$$\varphi_c = \frac{1}{2E}\sigma_c^2$$

and l_d is the characteristic length of the crack. To check the model, we study three cases of the behavior of the fracture under hydraulic injection. In the following, we summarize the main results of the study.

3.1 Case I. Cracking with Pre-existing Fractures

Here we aim to study how a pre-existing fracture (e.g., in natural rocks as shales) affects the crack path compared to the case without them (Fig. 12). Initially both cases evolve equally, until the displacement front reaches the surroundings of the pre-existing fracture (Fig. 13). Henceforth, it is clearly seen how the existence of fractures affects the path propagation. We can appreciate how the displacement front (driven by pressure injection) collides the pre-existing fracture and causes the maximum displacement gradient to change direction with respect to the midpoint of the tip of the fracture. Due to the existence of the fracture and the stress concentration, it causes the curvature of the path.

Once the fractures are connected (Fig. 14) there is a pressure drop in the lower fracture together with the pressure increase in the pre-existing fracture which causes the deviation of the path of the superior fracture.

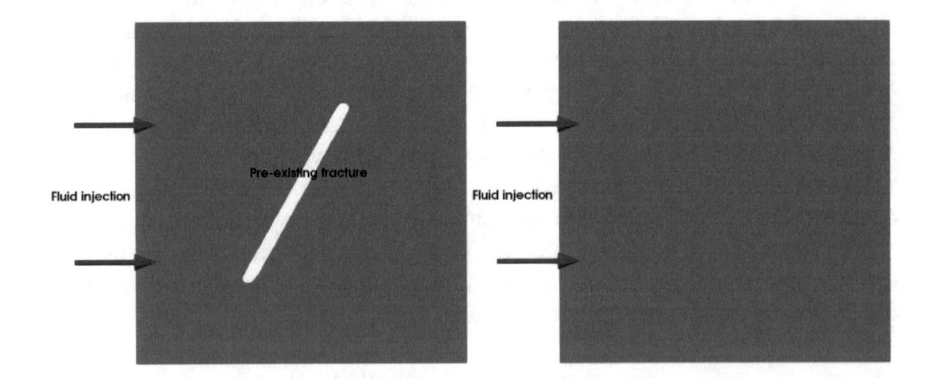

Fig. 12 Set up case with pre-existing fracture and without (same initial and boundary conditions, except pre-existing crack)

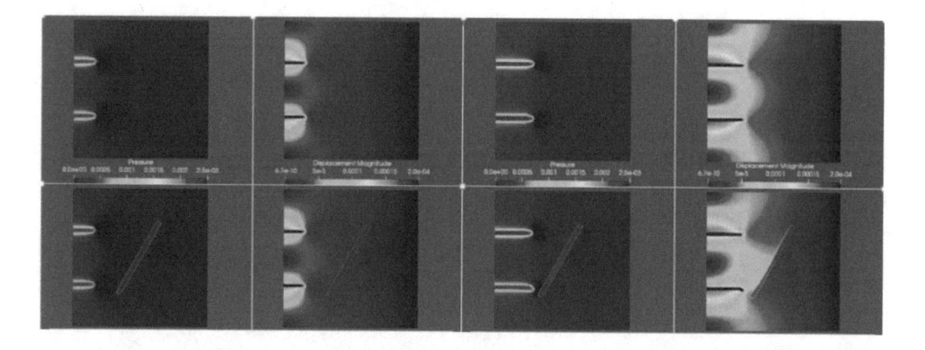

Fig. 13 Pressure and displacement fields *in* $t = 0+$ and right before collision: pre-existing fracture (*bottom*) and without initial crack (*top*)

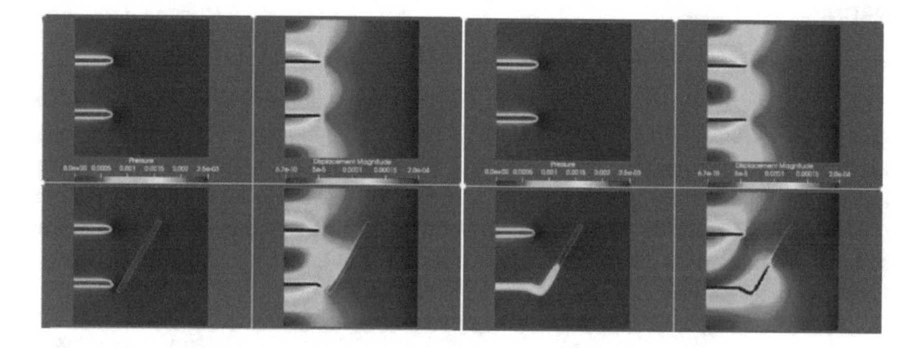

Fig. 14 Pressure and displacement right before and after collision

3.2 Case II. Effect of the Distance Between Fractures on Propagation

Very often we find the case of cracking with multiple injection points and the question arises of the optimum distance between points. In the following, we briefly show how the propagation of cracks is affected by the initial distance of two channels of injection, with or without previous natural fractures. We used the same boundary and initial conditions that in the previous section but changing the distance between fractures. In Fig. 15 (left) we show several cases in the scenario without natural fractures and we note that for large distances (blue) the cracks propagate independently in a parallel way but below a critical distance (red case) the symmetry is lost and the cracks bifurcate because one crack begins to influence to the other. We find this crack bifurcation problem interesting and a more systematic work is needed to relate the critical distance to mechanical properties of the system and to study the branching phenomena in depth.

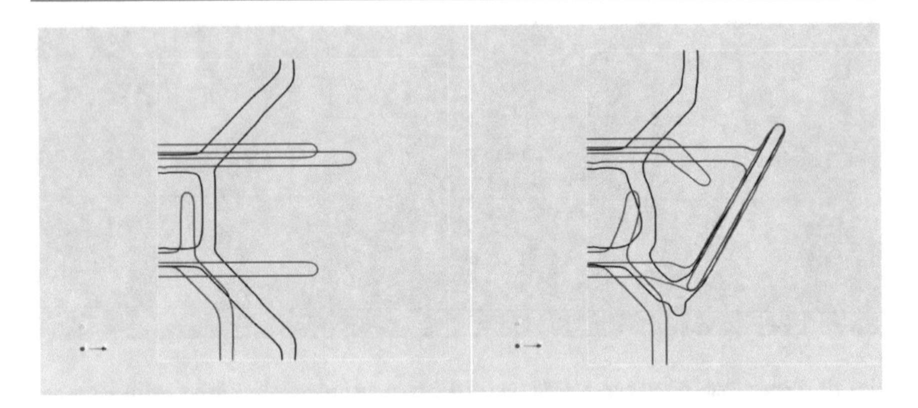

Fig. 15 Comparison of the final states of the fracture trajectory caused by injection without (left) and with (right) pre-existing fracture. Blue, greater initial distance, yellow, smaller distance in the injection

In Fig. 15 (right) we show the branching when a natural fracture is present. To specify an initial fracture involves a lot of other new geometrical parameters, e.g., the inclination, the length, the width, etc., which increase the number of scenarios but not the problem complexity. In the example, we have chosen only one possible choice and we observe that the length of the natural fracture (in particular the distance in the normal direction to the propagation) in relation to the initial distance between injection points is an important parameter.

We remark that such detailed treatment cannot be extended to be used in real oil fields. Nevertheless, it is useful to gain insight and develop subgrid models. For example, in a naturally fractured media, with hundreds or thousands of fractures, it is better to perform a statistical analysis using averaging techniques. However, the detailed models presented here are valuable in order to upscale the phenomena from the crack (or few cracks) scale to the real reservoir scenario with hundreds of fractures. This upscaling is in the same spirit as the permeability studies at pore scale give us a lot of information that is injected in the coarser Darcy (or porous media) scale.

3.3 Case III. Preferred Cracking Directionality

We now analyze the propagation of an initial small crack hydraulic fracture under different principal stress directions. It is reasonable to assume that in an isotropic and homogeneous medium, the principal stresses determine the final propagation and orientation of the fracture. But the question arises as to how long the orientation of a pre-existing fracture affects propagation and how it influences the orientation of the generated fracture.

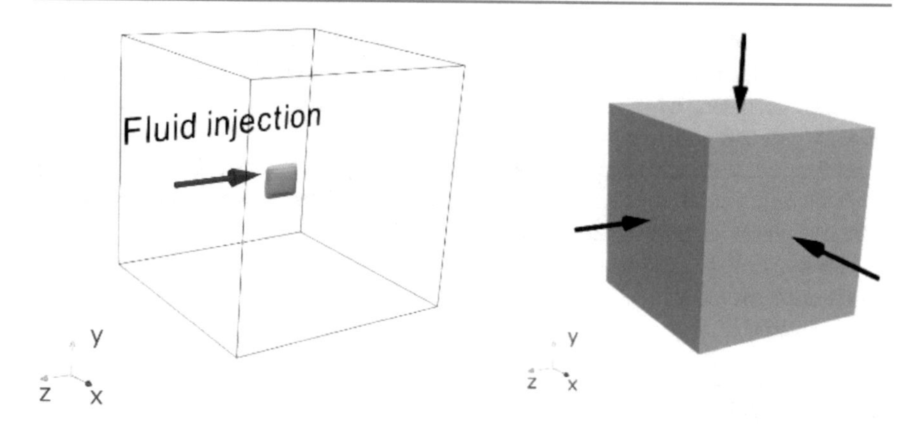

Fig. 16 Initial condition in the tree 3D cases (left). Boundary conditions (right): where hidden faces are set to zero displacement and visible faces are under compression (with different force modules in each case)

The proposed scenario is a small initial fracture where the injection is performed in the YZ plane that passes through the center of the domain (Fig. 16).

Next, we simulate how the fracture propagates under different boundary conditions. This propagation always occurs in the direction of maximum effort and the opening in the direction of minimum effort (Fig. 17), since the fracture opens in the direction of minimum energy and generates a region around the tip with a stress concentration until the breaking energy is reached.

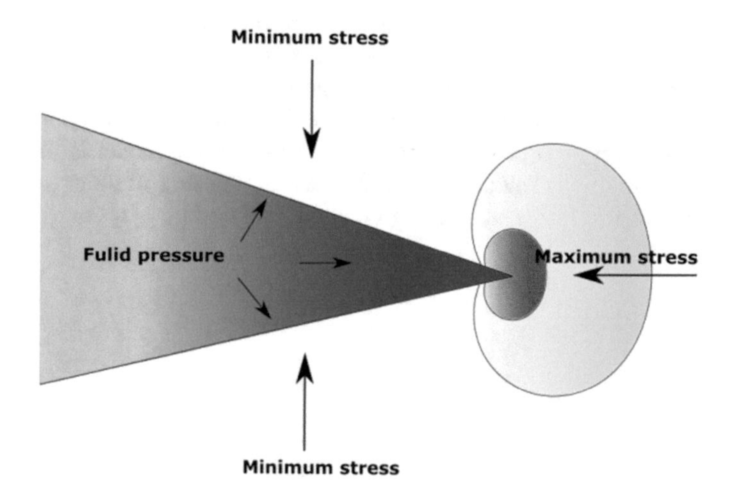

Fig. 17 Sketch of stress diagram during hydraulic fracturing

The procedure is the following: keeping the initial crack at a fixed position and orientation, three scenarios have been simulated with different boundary conditions. For each scenario, we choose the x-stress (respectively, y and z-stresses) in the boundary as the minimum of the three ones.

In Fig. 18, we show several snapshots during crack evolution. For each snapshot three cases are shown with different clippings in order to better appreciate the different planes of propagation of the fracture.

Finally, in Fig. 19 it is shown, just as example, the propagation in the case for minimum stress in z.

4 Conclusions

We have presented a brief and illustrative review of the advantages and suitability of the phase field method to address a wide range of problems in many fields, emphasizing the elements that mark the differences between applications. We think that the phase field theory is mature and able to cope with problems ranging from the simple spinodal decomposition, nucleation, and growth of two, three, and more phases in microstructure of alloys and condensed phases, wetting dynamics in surfaces, mesoscopic evolution of drops, ganglia or bubbles in porous media, macroscopic multiphasic flows, anisotropic growth and crystallization of solid phases, electrochemical reactions and dynamics in electrodes or metastatic tumor cells migration. It is being used on a daily basis in different fields, at least in a qualitative level, and it is an invaluable tool to understand the coupled physical processes in many systems. Many of the examples that we have shown in the review are part of our daily work along the last years using own in-house codes. The last example on the application of phase field theory to the propagation of hydraulic fractures in solids is the most interesting one from the oil-and-gas industry's point of view. We have delved into this example in greater detail as it combines phase field theory with flow in porous media and rock mechanics. As preliminary results we find that it reproduces well the process but no comparison to detailed experiments has been done yet. We think that the model is in the right direction and we find it suitable as a tool in the formulation of either reduced or sub-grid models for further upscaling to real oil field scale.

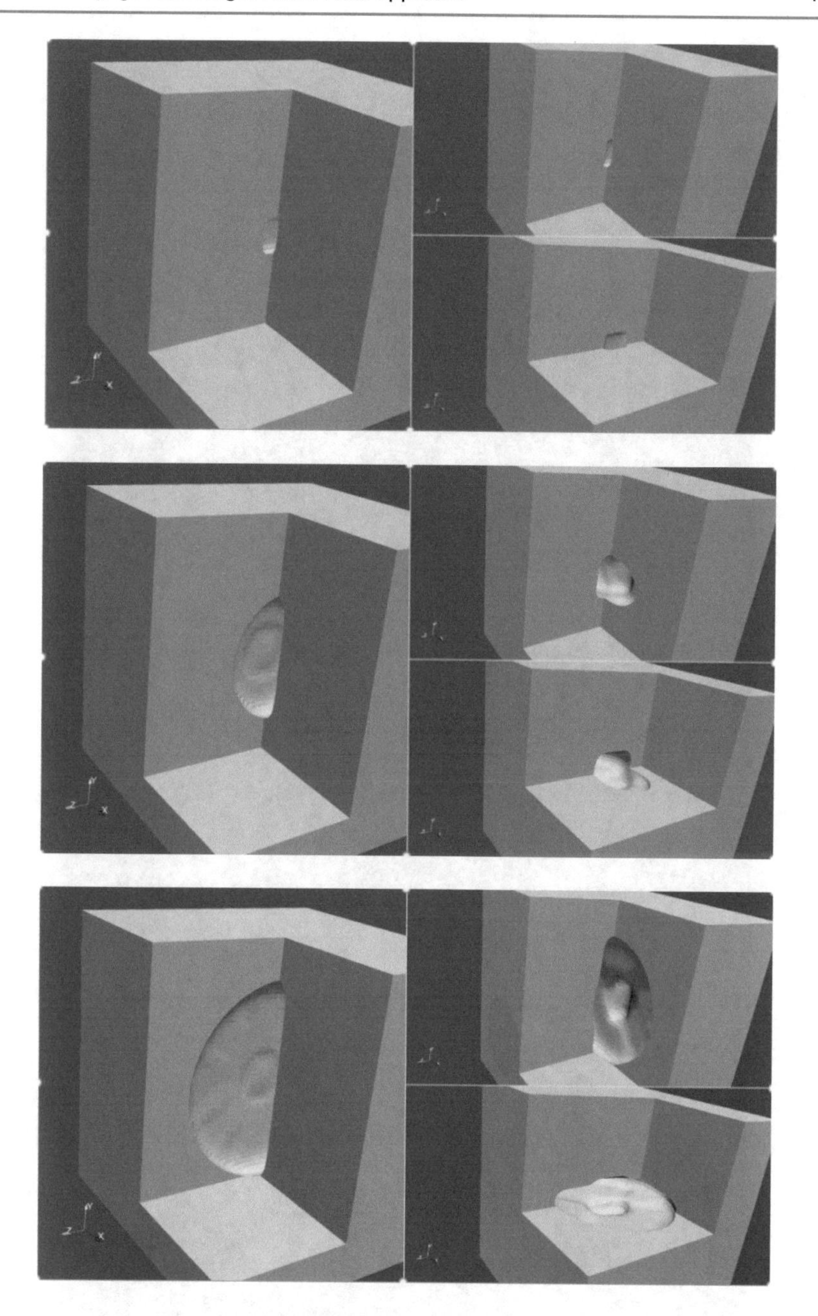

Fig. 18 Crack evolution at three times (from top to bottom): minimum stress in x-direction (left), minimum stress in z-direction (upper right), minimum stress in y-direction (lower right)

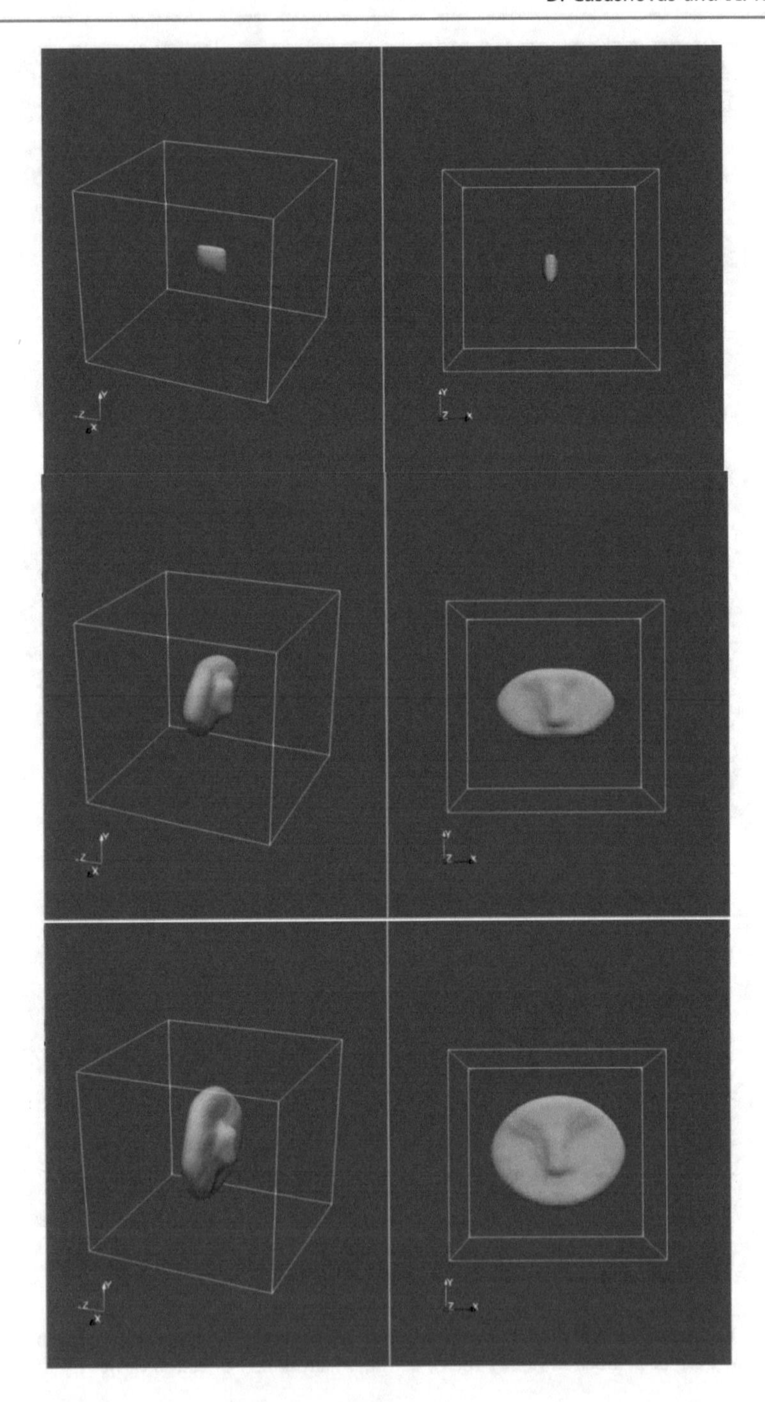

Fig. 19 Evolution of the phase field (fracture) in three instants of time in the case of the minimum effort in the z-direction

References

1. Anderson, D., Edwards, D., Raymond, C.: Asymptotic results for a barrier potential model, "Phase Field Formulation for Microstructure Evolution in Oxide Ceramics." In: Edwards, D. (ed.) 29th Annual Workshop on Mathematical Problems In Industry (2013)
2. Bai, P., Bazant, M.Z.: Charge transfer kinetics at the solid-solid interface in porous electrodes. Nat. Commun. **5**, 3585 (2014)
3. Boettinger, W.J., Warren, J.A., Beckermann, C., Karma, A.: Phase-field simulation of solidification. Annu. Rev. Mater. Res. **32**, 163–194 (2002)
4. Boyer, F., Duval, F., Intro, C.: Cahn-Hilliard/Navier-Stokes model for the simulation of three-phase flows. Nonlinear Mech. Multiph. Flow Porous Media, 1–32 (2008)
5. Boyer, F., Lapuerta, C.: Study of a three component Cahn-Hilliard flow model. Optimize **1** (2004)
6. Cahn, J.W., Hilliard, J.E.: Free energy of a nonuniform system. I. Interfacial free energy. J. Chem. Phys. **28**, 258–267 (1958)
7. Cogswell, D.A.: Quantitative phase-field modeling of dendritic electrodeposition. Phsy. Rev. **011301**, 1–5 (2015)
8. Cogswell, D.A., Carter, W.C.: Thermodynamic phase-field model for microstructure with multiple components and phases: the possibility of metastable phases. Phys. Rev. E Stat. Nonlinear Soft Matter Phys. **83** (2011)
9. Courant R., Hilbert, D.: Methods of Mathematical Physics, vol 1, Chap. 4, 164–274. Interscience Publishers Inc. (1953)
10. Desplat, J.C., Pagonabarraga, I., Bladon, P.: Ludwig: a parallel Lattice-Boltzmann code for complex fluids. Comput. Phys. Commun. **134**, 273–290 (2001)
11. Donaldson, A.A., Kirpalani, D., Macchi, A.: Diffuse interface tracking of immiscible fluids: improving phase continuity through free energy density selection. Int. J. Multiph. Flow **37**, 777–787 (2011)
12. Henry, H.: Study of the branching instability using a phase field model of in-plane crack propagation. Europhys. Lett. **83** (2008)
13. Hohenberg, P., Walter, K.: Inhomogeneous electron gas. Phys. Rev. **136**(3B) (1964). https://doi.org/10.1103/physrev.136.b864
14. https://en.wikipedia.org/wiki/Spinodal_decomposition
15. Kobayashi, R.: Modeling and numerical simulations of dendritic crystal growth. Phys. D Nonlinear Phenom. **63**, 410–423 (1993)
16. Ledesma-Aguilar, R.: Hydrophobicity in capillary flows. PhD dissertation thesis UB (2009)
17. McFadden, G.B., Wheeler, A.A., Braun, R.J., Coriell, S.R., Sekerka, R.F.: Phase-field models for anisotropic interfaces. Phys. Rev. E **48**, 2016–2024 (1993)
18. Miehe, C., Mauthe, S.: Phase field modeling of fracture in multi-physics problems. Part III. Crack driving forces in hydro-poro-elasticity and hydraulic fracturing of fluid-saturated porous media. Comput. Methods Appl. Mechan. Eng. (2016)
19. Miehe, C. Welschinger,, F., Hofacker, M.: Thermodynamically consistent phase-field models of fracture: variational principles and multi-field FE implementations. Int. J. Numer. Methods Eng., 1273–1311 (2010)
20. Moure, A., Gomez, H.: Phase-field model of cellular migration: three-dimensional simulations in fibrous networks. Comput. Methods Appl. Mech. Eng. **320**, 162–197 (2017)
21. Moure, A., Gomez, H.: Three-dimensional simulation of obstacle-mediated chemotaxis. Biomech. Model. Mechanobiol. **17**, 1243–1268 (2018)
22. Steinbach, I.: Phase-field model for microstructure evolution at the mesoscopic scale. Annu. Rev. Mater. Res. **43**, 89–107 (2013)

23. Wang, S.L., et al.: Thermodynamically-consistent phase-field models for solidification. Phys. D Nonlinear Phenom. **69**, 189–200 (1993)
24. Witelski, T.: Linear stability analysis of phase field models, "Phase Field Formulation for Microstructure Evolution in Oxide Ceramics." In: Edwards, D. (ed.) 29th Annual Workshop on Mathematical Problems in Industry (2013)

Phase Space Learning with Neural Networks

Jaime López García and Ángel Rivero

Abstract

This work proposes an autoencoder neural network as a non-linear generalization of projection-based methods for solving Partial Differential Equations (PDEs). The proposed deep learning architecture presented is capable of generating the dynamics of PDEs by integrating them completely in a very reduced latent space without intermediate reconstructions, to then decode the latent solution back to the original space. The learned latent trajectories are represented and their physical plausibility is analysed. It is shown the reliability of properly regularized neural networks to learn the global characteristics of a dynamical system's phase space from the sample data of a single path, as well as its ability to predict unseen bifurcations.

1 Introduction

Despite the constant improvement in computing power, the intrinsic degrees of freedom present in some regimes of complex non-linear model solutions, such as turbulent flows, make them unsuitable for control applications, or extremely difficult to solve for complex geometries. Recently, there is ongoing research to find more reliable and compressed order reduction methods out of complex physical models in different areas, such as computer graphics, control engineering, computational biology and other simulation disciplines.

In the last few years, *deep autoencoder* architectures have proven successful in compressing highly complex non-linear data distributions, which arise in image and

J. López García · Á. Rivero (✉)
Repsol Technology Center, Madrid, Spain
e-mail: angel.rivero@repsol.com

J. López García
e-mail: jaime.lopez.garcia@repsol.com

P. Quintela Estévez et al. (eds.), *Advances on Links Between Mathematics and Industry*, SxI - Springer for Innovation / SxI - Springer per l'Innovazione 15,
https://doi.org/10.1007/978-3-030-59223-3_8

audio classification scenarios. Therefore, it is achieved a compact latent representation that encodes data variation factors in a reduced parametrization, allowing the generation of new samples that are indistinguishable from the real ones.

There is an emerging interest in applying the previous ideas to the domain of simulation, which is enhanced by the achievements made so far in recent works, [1,2]. For this reason, this paper proposes the use of autoencoders as a non-linear generalization of projection-based order reduction methods, such as Proper Orthogonal Decomposition (POD).

We show, for a set of well known non-linear Partial Differential Equations (PDEs), how deep learning models are able to find a compact basis that encodes the dynamics of the original problem (1), into a small set of Ordinary Differential Equations (ODEs), in the same way that a linear wave equation can be reduced to a set of uncoupled harmonic oscillators.

$$\begin{cases} \dfrac{\partial u}{\partial t} = F\left(u, \dfrac{\partial u}{\partial x}, \ldots, \dfrac{\partial^n u}{\partial x^n}\right). \\ \text{Boundary Conditions (BC)} \\ \text{Initial Conditions (IC)} \end{cases} \tag{1}$$

In order to understand the capacity of neural networks to learn the global dynamics of the reduced system, we first work with phase diagrams of 2D and 3D dynamical systems. It is shown how the network is capable of learning different representative behaviours, such as dissipative dynamics, limit cycles or chaoticity. These results are then applied to make the autoencoder network learn the dynamics of reduced PDEs, proving that they can be completely solved in the hidden states of a neural network.

The great compression capacity provided by neural networks makes it possible to encode higher non-linear PDEs on 3D basis, where their dynamic behaviour can be represented and their physical coherence can be easily verified.

The contribution of this work is structured in two parts. In the first part, we study classical dynamic systems representative of different phenomena. In this way, it is demonstrated that neural networks properly regularized with the loss function given in [3], can learn robust approximations of a dynamic system phase space, revealing similar responses to external driving forces and identifying the bifurcations that occur due to the system parameters variation. In the second part, autoencoder neural networks are introduced as a natural non-linear extension of projection-based reduction order methods. Subsequently, it is shown how this model allows to reduce the derivative of a non-linear PDE into a compact representation, which can be integrated as a small system of ODEs. The solution can be transformed back to the original representation by using the decoded function learned by the autoencoder.

First, it is necessary to evaluate and understand the ability of a network to extrapolate and generalize the dynamics learned from a discrete set of solved trajectories; that is, to learn the time derivative function, $f(x, t)$, of the solution x, of the dynamical system being studied. We will use $\hat{x}$ to represent the coordinates in the replicated system. Next, we want to study the capacity of a latent variable neural network model to encode a high-dimensional phase space into a reduced one, in which we can calculate latent trajectories, while learning a mapping that encodes them to the original space, in a manner similar to a non-linear POD. The coordinates in the reduced replicated system are noted as h.

Hence,the search for a valid neural network based on a reduced order model for (1) can be carried out in two phases according to the following scheme:

$$\text{Phase I:} \quad \begin{cases} \dot{x} = f(x, t) \\ \text{IC} \end{cases} \Longrightarrow \begin{cases} \dot{\hat{x}} = \hat{f}(\hat{x}, t) \\ \text{IC} \end{cases} \tag{2}$$

$$\text{Phase II:} \quad \begin{cases} \partial_t u = f(u, \partial_x u, \ldots, \partial_{x^n} u, t) \\ \text{BC} \\ \text{IC} \end{cases} \Longrightarrow \begin{cases} \dot{\hat{h}} = f_h(h, t) \\ \text{IC} \end{cases} \tag{3}$$

In the following section, there is a collection of related work, that has served as a basis for the elaboration of this research. Following the structure indicated above, two main sections are differentiated. Each section is divided into the corresponding methods and results. The work is concluded with a brief conclusion about the results obtained.

2 Related Work

Regarding to the first part of the work, learning the function $f(x, t)$ leads to a certain error ϵ, such that $\hat{f}(x, t) = f(x, t) + \epsilon(x, t)$. For a given domain, it can be proved [4,5] that this provides an upper bound to the error of the predicted trajectories. This, in combination with the *universal approximation theorem* for neural networks [6], leads to results that establish the capability of recurrent neural networks to arbitrarily approximate any dynamical system in a bounded domain [4,5].

About the learning of general dynamics with neural networks, [7] shows how *recurrent-feed-forward* networks can learn autonomous custom phase space portraits with different number and types of attractors. Recently, [8] has extended this approach to non-autonomous systems, coupling networks that were trained separately, to accurately reproduce the dynamics of coupled systems.

The ability to learn chaotic maps is shown in [9], which extends to the study of learning chaotic dynamical systems in [10]. In addition, [11] uses *echo-state networks* to learn higher dimensional chaotic attractors, that are present in the *Kuramoto–Sivashinsky* equation.

Long-term dynamic prediction is featured in [12], where a contractive loss is used to constraint the spectral radius of the jacobian matrix of the learned derivative function, thought to be linked with the unstability of the system.

A very interesting approach to get a better approximation of the system derivative was suggested in [3], by building a loss function out of numerical multi-step schemes.

In the second part of the work, with the aim of synthesizing a reduced order model, as shown in the diagram (3), we face a dynamics identification problem, since the functional form of the latent dynamics is not known. In this area, [13] achieves great results using a dictionary sparse regression approach. It takes advantage of the fact that most dynamical systems present a sparse representation on a polynomial basis.

A POD neural network hybrid approach was introduced in [14], where Long Short-Term Memory (LSTM) networks are fed with the *k-component* POD solution projection, to learn the reduced dynamics. Furthermore, in [15], high-dimensional systems were solved entirely in the state-space of an LSTM network.

The most significant results achieved so far are shown in [2], where convolution autoencoders coupled with LSTM in their latent space, are used to learn reduced models.

3 Phase Space Learning

This first part of the document corresponds to the steps described in the diagram in (2).

3.1 Methods

Following the scheme given in (2), it is necessary to evaluate the capability of neural networks to learn phase space dynamics in bounded regions from a small set of trajectories, in most cases only one.

In the first place, it is required to define the metrics used to evaluate the quality of the replicated dynamics.

The Mean Square Error (MSE) of the differences between the ground truth and the predicted path is not a significant measure by itself, as will be explained later. We want to study the capacity of the network to preserve general qualitative features of the phase space, beyond the Least Square Errors (L2) obtained.

We will take into account the following validation criteria:

1. Robustness of the learned attractor to different initial conditions.
2. Ability to extrapolate unseen regions of phase space.
3. Correct response to external perturbations.
4. Capacity to capture bifurcations in the dynamics.

Once the evaluation criteria are defined, our goal is to obtain a machine learning algorithm to approximate the derivative, f, of an arbitrary dynamical system. For this, it is necessary to establish the problem and select an architecture.

The machine learning problem of replicating the system in (2), is a probabilistic one. We are trying to find from the set of all $\hat{f}(x, \theta)$ that our model can represent, the one most similar to f. Thus, we are not modelling a deterministic transition $X_{t+1}(X_t)$ such as the one computed by solving an ODE, but a probabilistic one $P(X_{t+1}|X_t)$, where the unique deterministic transitions arising from the solution of first-order ODEs, are replaced by the $markovian \quad property$ of $markov \quad processes$, present in processes where the conditional probability distribution of future states of the process depends only upon the present state.

We are looking for a parametrized distribution $\hat{P}(f|X,\theta)$ that can be used to sample $\hat{P}(X_t \mid X_{t-1})$ with a numerical scheme. The machine learning problem is stated by requiring , $\hat{P}$, to be as close as possible to $P(f|X)$. This means that we aim to minimize the Kullback–Leibler divergence (KL) between the two distributions, which leads to the following cost function:

$$\underset{\theta}{\operatorname{argmin}}\, KL(\theta, X) = \underset{\theta}{\operatorname{argmax}}\, \frac{1}{N}\sum_{k=1}^{n} \log(\hat{P}(\hat{f} \mid X(t_k), \theta)), \tag{4}$$

Since f is a non-observable variable, we must use it implicitly by modelling the $P(X_t \mid X_t \dots X_{t-k}, f_t \dots f_{t-k})$ distribution. Employing a *multi-step* scheme (5), as in [3], the cost function (6) is obtained.

$$y_{t+s} = \sum_{i=0}^{s-1} \alpha_i y_{t+i} + \sum_{i=0}^{s} \beta_i f(y_{t+i}), \tag{5}$$

$$\underset{\theta}{\operatorname{argmax}}\, \mathcal{L}(X,\theta) = \underset{\theta}{\operatorname{argmax}} \frac{1}{N}\sum_{t=0}^{N} \log(P(X_{t+k} \mid f_t \dots f_{t+k}, X_t \dots X_{t+k-1})). \tag{6}$$

Finally, assuming a gaussian distribution with constant variance $P(X_{t+k} \mid f_t \dots f_{t+k}, X_t \dots X_{t+k-1}) = \mathcal{N}(\mu, \sigma)$,where $\mu = \mu(f_t \dots f_{t+k}, X_t \dots X_{t+k-1})$, we arrive at

$$\mathcal{L}_{ms}(X,\theta) = \frac{1}{N}\sum_{\forall t \in T} \left\| X_{t+s} - \sum_{i=0}^{s-1} \alpha_i X_{t+i} - \sum_{i=0}^{s} \beta_i \hat{f}(X_{t+i},\theta) \right\|_2^2 . \tag{7}$$

We employ a fully connected feed-forward architecture, as in [3,7,8,12], instead of a recurrent neural network. This is because dynamical systems can be understood as the differential deterministic equivalent of Markov processes, where $P(X_t \mid X_{t-1}, X_{t-2}, \dots) = P(X_t \mid X_{t-1})$.

As mentioned in the previous section, all the studied systems have a derivative $f(x,t)$ with a sparse representation on a polynomial basis [13]. This concept is integrated into our architecture by using an encoder scheme, along with a linear regressor, which will attempt to represent the monomials present in the system, as depicted in Fig. 1.

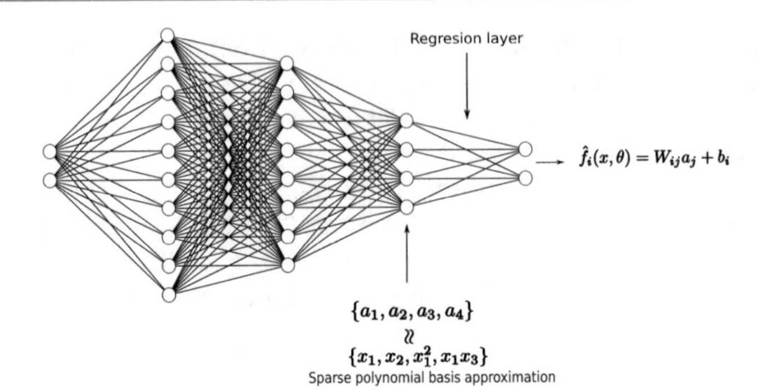

Fig. 1 Architecture. The last hidden layer enforces with a number of neurons equal to the number of different monomials in the system, the prior knowledge that the system is represented in a sparse polynomial basis

Learnt system error decomposition

The assessment of the learnt system error, boils down to the comparison between solutions of the two systems (8) with the same initial condition x_0.

$$\begin{cases} \dot{\hat{x}} = \hat{f}(\hat{x}) \\ \hat{x}_{t_0} = x_0 \end{cases} \qquad \begin{cases} \dot{x} = f(x) \\ x_{t_0} = x_0. \end{cases} \tag{8}$$

For this purpose, a new variable is defined that quantifies the discrepancy of solutions, $z = \hat{x} - x$. Without loss of generalization, we can write $\hat{f}(x) = f(x) + \epsilon_d(x)$, where $\epsilon_d(x)$ stands for the error made at each point by the network in the calculation of the derivative.

Replacing $z = \hat{x} - x$ in (8) leads to the following system:

$$\begin{cases} \dot{z} = \hat{f}(x+z) - f(x) \\ z_{t_0} = 0. \end{cases} \tag{9}$$

At the beginning of the trajectory z will be small, so we can do a power series expansion of $\hat{f}$ which will be valid for small times and will allow us to have a better understanding of the different sources or error:

$$\dot{z} = \hat{f}(x) + \nabla\hat{f}(x)\cdot z - f(x) = \epsilon_d(x) + (\nabla f(x) + \nabla\epsilon_d(x))\cdot z. \tag{10}$$

Each term in Eq. (10) has a straightforward interpretability, explicit in (11).

$$\begin{aligned} \dot{z}_l &= \epsilon_d(x) \\ \dot{z}_n &= \nabla f(x)\cdot z_n \\ \dot{z}_s &= \nabla\epsilon_d(x)\cdot z_s. \end{aligned} \tag{11}$$

The three equations in (11) account for the different growth factors of the difference between the network prediction and the ground truth trajectory.

The local error, z_l, assess the local discrepancy that stems from the difference in the attained derivative.

The "natural" term, z_n, accounts for the expected divergence of close trajectories embedded in the dynamics of the system itself. In chaotic systems, the hypersensitivity to different initial conditions is going to pull trajectories apart, even if the prediction of the network is arbitrarily good. This is the main reason why we should not focus on the mean squared difference of the predicted and real trajectories to validate the quality of the prediction.

We call z_s the stability error because it is of the same nature as the one that appears in the stability analysis of numerical schemes, as finite differences. It states important properties about the approximation of $f(x)$ that the network should achieve. It clearly indicates that penalizing the difference $\|\hat{f}(x) - f(x)\|_2 = \|\epsilon_d\|_2$ is not enough. Additionally, it is desirable that the eigenvalues of $\nabla\epsilon_d$ are negative or small, so the network errors z_s are dissipative. In practice, this means the implementation of a kind of regularization penalty, which smooths the features that the network learns.

This is especially important in systems that exhibit conservative dynamics, which require very fine-tuning of learned parameters. It is well known that *encoder–decoder* bottleneck architectures tend to learn contractive mappings that flow into the learnt manifold, this is, they locally behave as a linear dynamical system with $\nabla\epsilon_d(x)$ matrix presenting all negative eigenvalues, (or close to zero eigenvalues if we consider the associated differences equation). This character might be enhanced by certain architecture choices, such as in *contractive autoencoders* [16] or *denoising autoencoders* [17].

Training

For all the systems studied, we compared Adam–Moulton and Backward Differentiation Formula (BDF) schemes, with the former showing the best performance in extrapolating the dynamics of unseen regions in the phase space.

We found no evidence of the results getting worse as the order of the schemes is increased; on the contrary, the greater the order, the better the replication of the dynamics, even for higher order, not $A - stable$ methods. Therefore, the BDF-6 scheme was chosen as the default scheme in the loss function.

We used Adam optimizer [18], with a decaying learning rate (starting in 1e-3), and a batch-size of 200. Also, Rectified Linear Unit (ReLU) [19] activation function is utilized in all experiments. Except for the Lorentz equation where a bigger capacity network was needed, for all the experiments in this section, a network comprised of three hidden layers of (40, 20, 10) neurons, plus a regression layer on top, was used. Network weights were initialized with a He Normal Initialization [20].

3.2 Results

Three different simple models have been studied. The harmonic oscillator as a linear system, the Duffing equation and the Lorentz equation as examples of non-linear systems. The results obtained for each of them are described below, starting with the oscillator.

Linear systems: harmonic oscillator

The general equation for the harmonic oscillator is a second-order linear differential equation (12), where m, γ_m and ω_{0m} are the mass, the damping coefficient and the stiffness of the oscillator, respectively.

$$m\ddot{x} + \gamma_m \dot{x} + \omega_{0m} x = 0. \tag{12}$$

To attain the phase space representation of the system, we divide the last equation by m, incorporating it in γ, ω_0, and then we transform it into a set of first order differential equations:

$$\begin{cases} \dot{y} = -\gamma y - \omega_0 x \\ \dot{x} = y \\ x(t_0) = x_0,\ y(t_0) = y_0. \end{cases} \tag{13}$$

The system given in (13) is particularly relevant because it constitutes the general representation of the first-order approximation to a wide range of (2D) non-linear systems in the neighbourhood of a critical point.

The best way to test the reliability of a dynamical system's reproducibility is to study its response to an external perturbation. Therefore, as indicated in [8], a sinusoidal signal of varying frequency has been added to a network trained with the homogeneous linear system (13), being the parameters $\gamma = 0.01$ and $\omega_0 = 1$.

Subsequently, it has been compared with the integrated numerical solution of the perturbed system. The resulting system corresponds to a forced damped harmonic oscillator:

$$\begin{cases} \dot{y} = -\gamma y - \omega_0 x + g(t) \\ \dot{x} = y \\ x(t_0) = x_0,\ y(t_0) = y_0. \end{cases} \tag{14}$$

The solution of a harmonically driven damped oscillator is a harmonic oscillation with an amplitude dependent on the frequency of the driving signal, showing a resonance peak near the natural frequency ω_0.

In Fig. 2, it is shown how similar the response of both systems is, presenting a resonance peak, (driving force frequency that causes the maximum amplitude), virtually at the same driving frequency.

Duffing equation

Duffing equation is a non-linear second-order differential equation, given by

$$\ddot{x} + \delta\dot{x} + \alpha x + \beta x^3 = \gamma \cos \omega t. \tag{15}$$

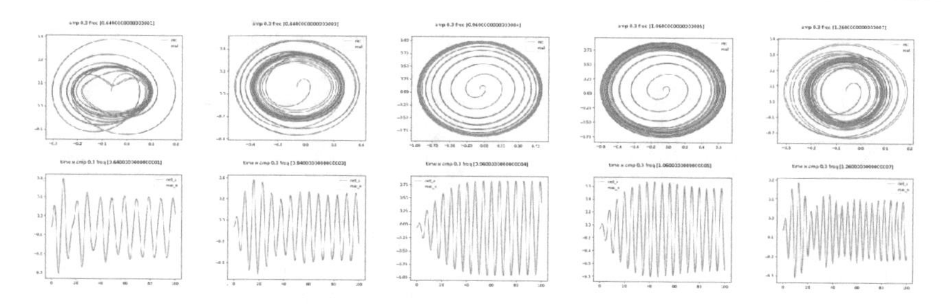

Fig. 2 Solution of a harmonically driven damped oscillator. Neural network performance in orange, numerical solution in blue

Equation (15), is suited for modelling oscillators with non-linear, odd restoring forces. Duffing equation can be expressed as a system of first-order differential equations, presented in (16).

$$\begin{cases} \dot{x} = y \\ \dot{y} = -\delta y - \alpha x - \beta x^3 + \gamma \cos \omega t \\ x(t_0) = x_0, \, y(t_0) = y_0. \end{cases} \tag{16}$$

By using the parameters $\gamma = 0, \alpha = -1, \beta = 1, \delta = 0.3$ and $\omega = 1.2$, the potential $V = \frac{1}{2}\alpha x^2 - \frac{1}{4}\beta x^4$, that generate the restoration force $-\nabla V = -\alpha x - \beta x^3$, is a two-well potential. Without any driving force, the system has two fixed point attractors, that correspond to a particle being trapped in one of the two wells, as shown in Fig. 3.

As the amplitude of the driving force, γ, increases, the particle is more capable of jumping from one well to another. Once a certain amplitude is exceeded, a series of *period-doubling* bifurcations take place, so the system's behaviour becomes chaotic.

Proceeding in an analogous manner to what was done with the damped-driven oscillator, we first train the network with a trajectory calculated solving the (16) system without the $\gamma \cos(\omega t)$ term. Once the network is trained, we add the driving term to the model output and we sweep across a range of γ values from 0 to 0.6, knowing that the bifurcation takes place near $\gamma = 0.3$.

As it can be seen in Fig. 3, both the neural network system and the original one, suffer a transition towards chaos at approximately the same value of γ. The trajectories of both systems differ considerably after some time. This is not an indicator of low accuracy, but rather the expected behaviour of a chaotic system, whose arbitrarily close starting points are carried away by positive eigenvalues of ∇f, as indicated in the error description part.

For γ values well above the bifurcation, the chaotic behaviour starts to decrease and the decorrelation times of both the network and the original systems increase. This is due to the fact that the *Maximum Characteristic Lyapunov Exponent* of the system decreases to a point where the dynamics are no longer chaotic.

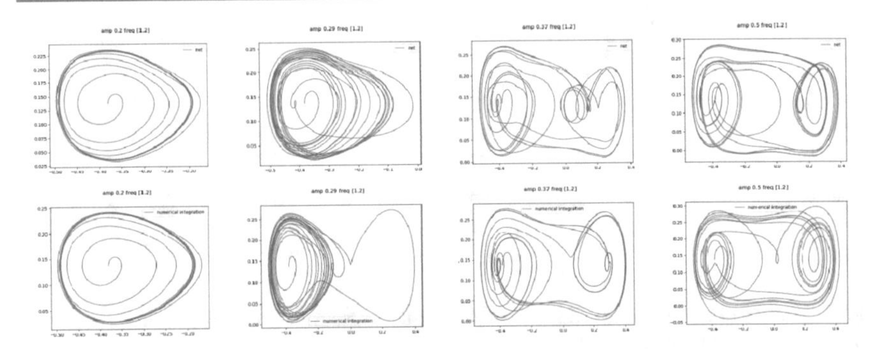

Fig. 3 Solution of an oscillator with non-linear odd restoring forces. The first row corresponds to a forced network trained on the homogeneous Duffing equation. The second one shows the ground truth forced system

Fig. 4 Lorentz's attractor

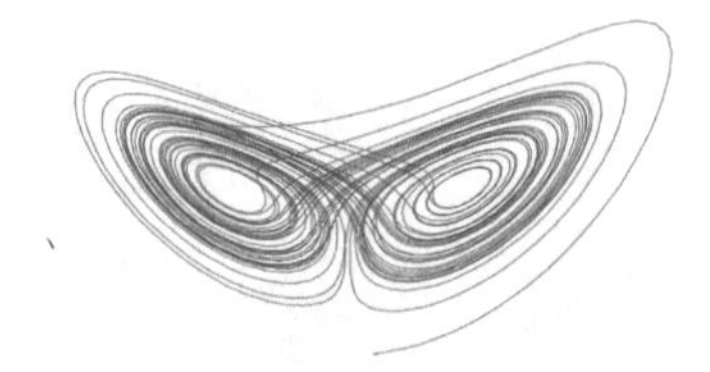

Lorenz equation

The Lorentz system (17) is comprised of a set of three non-linear ODEs and is frequently used as the paradigmatic system for the exemplification of chaotic solutions. It was developed as a simplified mathematical model for atmospheric convection, by truncating a series expansion of the *Navier–Stokes* equations, describing a two-dimensional fluid layer uniformly warmed from below and cooled from above.

$$\begin{cases} \dot{x} = \sigma(y - x) \\ \dot{y} = x(\rho - z) - y \\ \dot{z} = xy - \beta z \end{cases} \tag{17}$$

Using the values $\sigma = 10$, $\beta = 8/3$ and $\rho = 28$, the systems present chaotic solutions, where trajectories are pulled towards a fractal set that constitutes Lorentz's strange attractor, Fig. 4. This attractor presents global stability, in the sense that almost all points in the phase space are attracted to it. Once in the attractor, any point will spiral around one of the fixed points before shooting at the other symmetric one. The time it will spend on each of the wings, or which one the point will go to first, depends mostly on the initial conditions, which, although deterministic, are computationally unpredictable. This arises from the deep entanglement of the stable and unstable manifolds of the different fixed points [21]. Preserving this fine structure of phase space, needed to replicate chaotic behaviours, means that the network reconstruction of f should be much more accurate, hence, more capacity will be required.

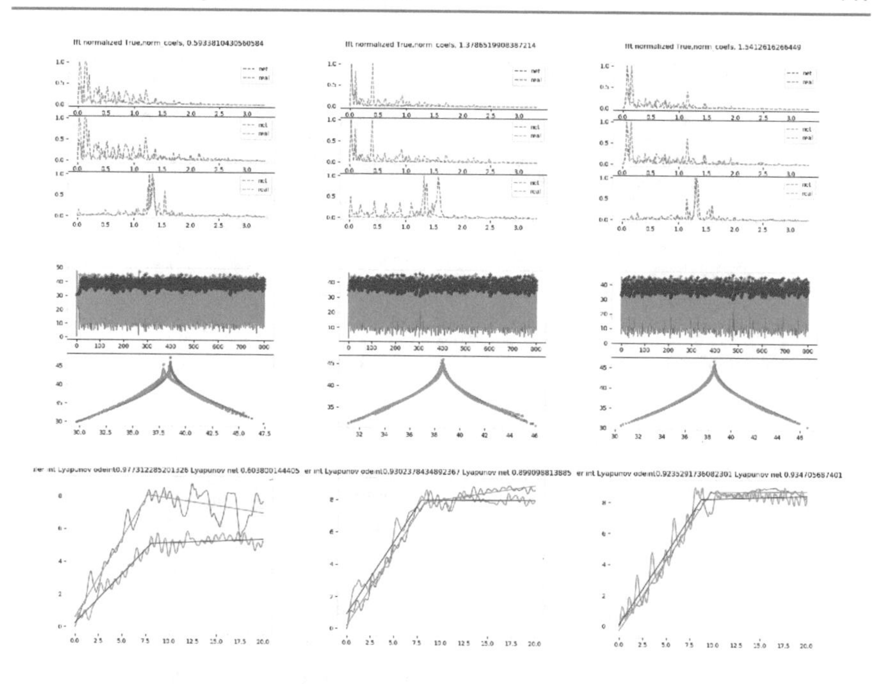

Fig. 5 Fourier transform, Lorenz maps and $MCLE$ estimations of a neural network with 3 hidden layers and 16 neurons (left column), 3 hidden layers and 30 neurons (middle column) and 2 hidden layers and 60 neurons (right column). Solution of the ODE in blue, approximated solution in orange. We can see how the spectral power composition of the solutions is better matched in the case of the higher capacity networks, while the small one fails at capturing higher frequencies.$MCLE$ is approximated by the slope of the average $log(distances)$ of initially close trajectories (the flattening part of the curve takes place after going through the Lorenz butterfly center, that acts as a bottleneck that $grinds$ initial correlations). We can see the close match of slopes between the curves of the approximated model and the original one

To obtain a reliable chaotic replication of the attractor, 100 neurons have been used in the first hidden layer. In this case, the Lorentz map, the Maximum Characteristic Lyapunov Exponent ($MCLE$) estimation and the Fourier transform of the trajectories agree on the chaotic nature of the learned attractor, as can be seen in Fig. 5.

By increasing the depth (number of hidden layers) of the network while maintaining the width (number of neurons), chaotic dynamics are achieved with a reduced number of parameters. This is because the number of regions that can be separated by a neural network grows exponentially with depth [22]. In addition, because the two fixed points are symmetrical, there are features that can be shared and do not need a separate parameterization, as it happens in a shallow architecture.

In Fig. 6, we use a neural network with 60 neurons in the first layer, 40 in the second, 4 in the third and 3 in the linear regression layer. It is shown how the global stability of the learnt attractor is preserved far from the training trajectory, promoting even further the perspective of deep neural networks as dynamical systems.

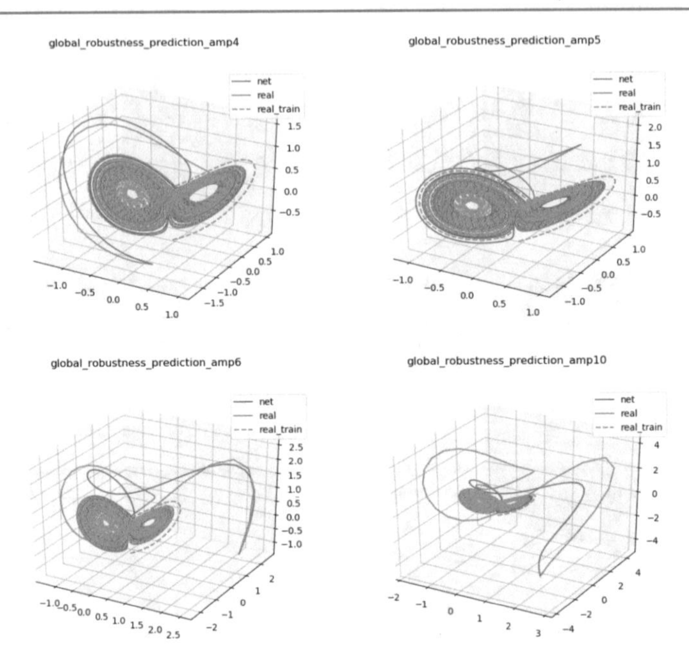

Fig. 6 Different trajectories for a neural network with 60 neurons and 2 hidden layers. Distant points are pulled back to the attractor. The training trajectory in green, the neural network results in blue and the test trajectory in orange

Another advantage of deep learning is the ability to learn bifurcations. By maintaining the value of σ and β parameters, but changing ρ, huge variations in system dynamics are observed. This structural modification is caused by changes in the stability of the fixed points and the appearance of new ones. The study of this phenomenon is included in the theory of bifurcation. Here is the behaviour for some significant values:

1. If $0 < \rho < 1$, the origin is the only equilibrium point, towards which all the orbits converge.
2. If $\rho = 1$, a *supercritical pitchfork* bifurcation occurs, the origin become unstable and a pair of stable symmetric fixed points appear.
3. If $\rho = \frac{\sigma(\sigma+\beta+3)}{\sigma-\beta-1} \approx 24.73$, the eigenvalues of both symmetric fixed points cross the complex plane and lose their stability through a subcritical *Hopf bifurcation* [23].

As done in [3], learning the bifurcations can be re-stated as learning the phase space of the system (18), where ρ is treated as an additional variable, whose fixed character is described by $\dot{\rho} = 0$.

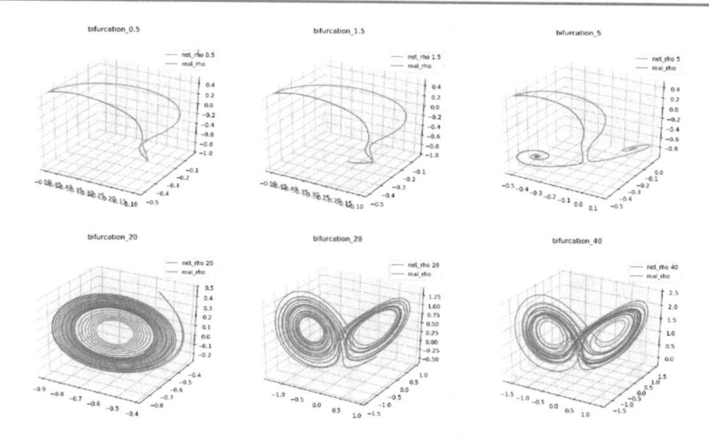

Fig. 7 Network predictions and numerical integration results for different values of ρ

$$\begin{cases} \dot{x} = \sigma(y - x) \\ \dot{y} = x(\rho - z) - y \\ \dot{z} = xy - \beta z \\ \dot{\rho} = 0 \\ \mathbf{x}(t_0) = \mathbf{x_0}, \rho(t_0) = \rho_0, \end{cases} \tag{18}$$

A network with 80, 60, 10 and 4 neurons in the hidden layers has been trained with $\rho \in \{1, 2, \ldots, 19, 20\}$, so that the network does not see data from orbits in regimes that correspond to $\rho < 1$ or $\rho > 24.7$. For the evaluation, the network is fed with the desired value of ρ and the conditioned trajectory is integrated as in the previous experiments. As shown in Fig. 7, the network is able to capture both non seen bifurcations at $\rho = 1$ and $\rho = 24.7$.

4 Latent Phase Space Learning

This second part of the document corresponds to the steps described in the diagram in 3. The methods and the training process described below, are schematically displayed in Fig. 8.

4.1 Methods

Once we have studied simpler models, in this second part of the work, we will move forward to deal with dynamical systems described by PDEs (1), with a large number of degrees of freedom. These can be reduced by searching for a suitable coordinate change, which better represents the structure of the latent phase space. The first training stage consists of an unsupervised learning stage, where we aim to capture the manifold where the trajectory exists, this is finding an initial basis to integrate the

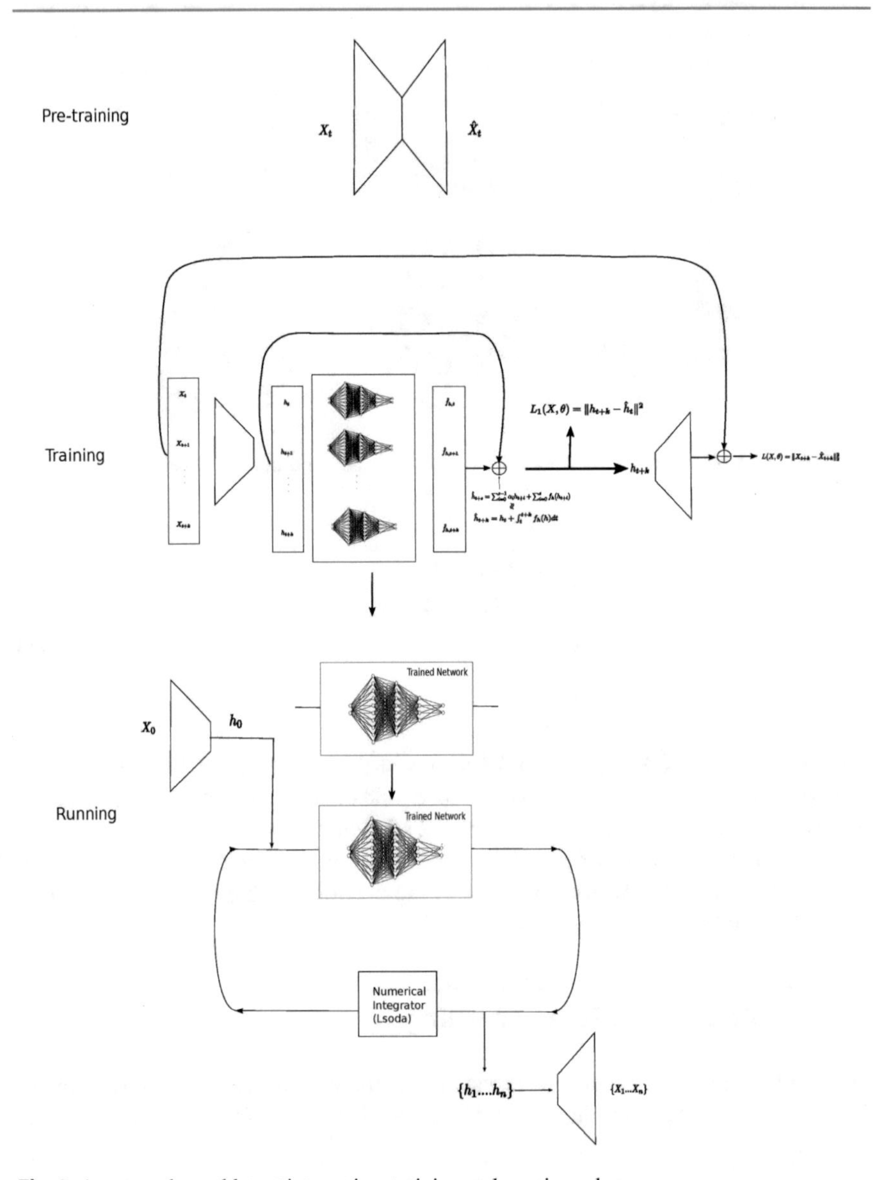

Fig. 8 Autoencoder and latent integration, training and running scheme

latent dynamical system, this step involves the minimization of the reconstruction loss term (19).

The probability distribution that we are modelling now is $P(X_{t+k}|X_t, \theta)$, and we want to factorize it into a projection to a reduced coordinate system $P(h_t|X_t)$ and a latent integration part $P(h_{t+k}|h_t)$, this factor would play the role of $P(x_{t+k}|x_t)$ in last section. Under the assumption of normal errors of constant variance, this factorization leads us to $P(X_{t+k}|X_t, \theta) = P(X_{t+k}|h_{t+k})P(h_{t+k}|h_{t...t+k-1}, f_t, X_t, \theta) = \mathcal{N}(\mu_1 = \hat{X}_{t+k}(h_{t+k}, \theta_{encoder}), \sigma)\mathcal{N}(\mu_2 = \hat{h}_{t+k}(\theta_{integration}, h_t(X_t, \theta_{decoder})), \sigma)$. With this

probability distribution, we can lay out the optimization problem in the same way as in part I, as it is shown in (20).

$$\begin{aligned} \underset{\theta}{\operatorname{argmax}}\, \mathcal{L}(X, \theta) &= \underset{\theta}{\operatorname{argmax}} \frac{1}{N} \sum \log(P(X_t|\hat{X}_t) \\ &= \underset{\theta}{\operatorname{argmin}}\, \frac{1}{N} \sum \left\| X_t - \hat{X}_t(h_t(\theta, X_t) \right\|^2 \end{aligned} \tag{19}$$

$$\begin{aligned} \underset{\theta}{\operatorname{argmax}}\, \mathcal{L}(X, \theta) = \underset{\theta}{\operatorname{argmax}}\, \frac{1}{N} \sum_{k=1}^{n} log(P(X_{t+k}|h_{t+k}) \\ P(h_{t+k}|h_{t+1} \ldots h_{t+k-1}, f_{ht+1} \ldots f_{ht+k}, X_{t+1} \ldots X_{t+k-1}))\,. \end{aligned} \tag{20}$$

Inserting the multi-step approximation of the latent derivative in the latent space of the autoencoder to model $P(h_t + k|h_t)$, we get the cost function (21). The autoencoder and the derivative parameters are denoted θ_{aut} and θ_{der}, respectively.

$$\begin{aligned} \mathcal{L}(X, \theta) = &\sum_{k=1}^{n} \left\| X_{t+k} - \hat{X}_{t+k}(h_{t+k}) \right\|^2 \\ &+ \left\| h_{t+k} - \sum_{i=0}^{k-1} \alpha_i h_{t+i}(X, \theta_{aut}) - \sum_{i=0}^{k} f_{ht+i}(h, \theta_{der}) \right\|^2 . \end{aligned} \tag{21}$$

Training

Training and evaluation are carried out by following these steps:

1. In a pre-training stage, X_t is reconstructed.
2. In the training stage, we insert the derivative network studied in the first part of this work that models $P(f_h|h)$.
3. In the evaluation stage, we unplug the trained derivative network, encode an initial condition $X_0 \rightarrow h_0$ and integrate the derivative network in the latent space. We later decode this integrated solution and recover the X representation of the trajectory.

For the autoencoder, it was used a 100 neuron and 9 hidden layers network with ReLU activations, except for the latent and regressor layers. It was used a width factor between layers of $3/4$, and a 60 ReLU, 40 ReLU, 20 linear network to model the latent derivative.

A denoising autoencoder [17] architecture was fundamental to learn useful robust features. A corrupted noise of amplitude 10^{-3} was injected in the normalized input data.

Adam optimization algorithm [18] was used in the *backpropagation* optimization to train the network, along with a 0.1 factor decaying learning rate every 200 epochs and a batch size of 200 samples. Smaller networks would be enough to successfully learn most of the problems, but for consistency's sake, the architecture was kept fixed.

5 Results

The PDE training data was acquired by solving different equations with the MATLAB spectral solver package `Chebfun`. We used 512 nodes and periodic boundary conditions in all cases, downsampled to 64 to train the network.

When solving equations using a spectral method with this resolution, some numerical artefacts may occur. This is the case of the formation of shock waves in the *Burgers equations*. High-frequency modes, which cannot be represented in discretization, give rise to phenomena like the Gibbs one. As will be seen below, this is not a problem for a neural network since it achieves a more compact and richer representation than the spectral one.

The integration time for each experiment was set in order to clearly manifest the characteristic dynamics given by each equation with every coefficient set to 1. Burger's equation integration time should capture the formation and damping of the shock wave. In the Korteweg–de-Vries equation, the time used captured 3–4 cycles of soliton (semi)periodic movement. In the Kuramoto–Sivashinsky equation, the time was set to ensure fully developed turbulent behaviour, where the intermittency characteristics of the turbulent patterns were shown.

Spatial domain bounds in all plots were scaled to $[-1, 1]$. The spatial domain for the Burger's equation was $[-8, 8]$,with an initial condition of $e^{-(x+2)^2}$ and integration time of 70. For the Korteweg–de-Vries equation, in the single soliton case, the spatial domain is $[-10, 10]$, with an initial condition $e^{-(x-7)^2/10}$ and an integration time of 12. For the multiple solition case, we used a domain of $[-20, 20]$ and an initial condition $sin(\frac{\pi x}{20})$. With Kuramoto–Sivashinsky equation, the spatial domain was $[0, 50]$ and the integration time was 250, with the initial condition $\cos(\frac{x}{8})(1 + \sin(\frac{x}{8}))$.

The temporal discretization used is problem-dependent. We observed that, in general, the fitted networks are able to preserve numerical stability with temporal deltas of greater magnitude than in the case of the spectral method integration, so, in every experiment, with the integration time fixed, we sought for the biggest dt that made the spectral method integration convergent and then downsampled it by

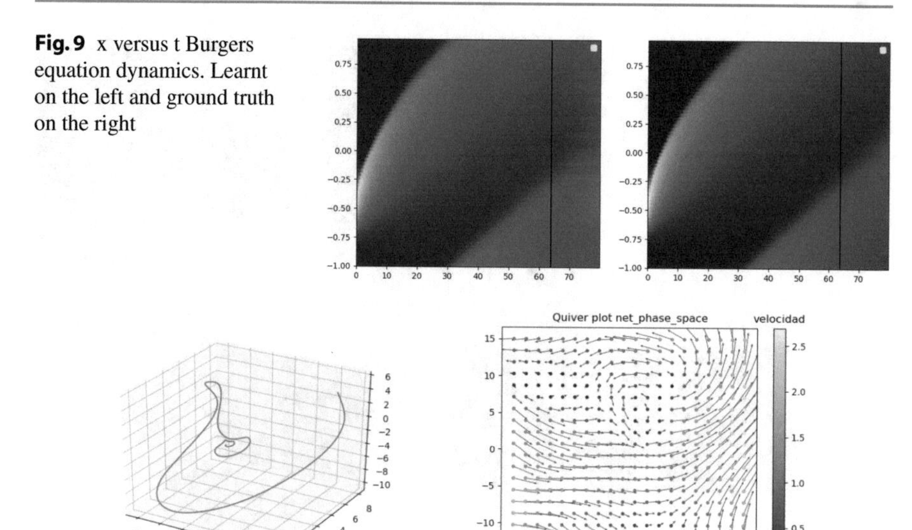

Fig. 9 x versus t Burgers equation dynamics. Learnt on the left and ground truth on the right

Fig. 10 Latent phase space and two principal component projection of the learnt latent phase space for the training and test time together

a factor of 2 to feed the network, we saved 25% of the last time samples for testing and used the rest for training.

Viscid Burgers

Burger's equation (22) is a simplification of the incompressible Navier–Stokes equation. It is useful to understand mathematical problems that arise in the Navier–Stokes equations under certain conditions, such as a limit of small viscosities, the formation of shock waves and so on. This is the simplest model that combines non-linear advection and diffusion. Eliminating the diffusion term, the equation exhibits discontinuities as the shock waves are formed. The presence of the diffusion term counteracts the effect of non-linearity, resulting in an equilibrium between the non-linear advection term and linear diffusion.

$$\frac{\partial u}{\partial t} + u\frac{\partial u}{\partial x} = \mu\frac{\partial^2 u}{\partial^2 x}. \tag{22}$$

In Fig. 9, both learnt and ground truth dynamics are shown for the Burgers equation. In Fig. 10, it is clear where the formation of the shock wave happens, as well as how the dissipative dynamics of the underlying system are correctly captured.

As mentioned above, once the training phase has been completed, the integration of the trajectories has been done using an off-the-shelf ODEs integrator (`lsoda` BDF-4, specifically). The latent derivative of the trained network is passed as the system derivative function, together with the initial condition, X_0, encoded into h_0 coordinates through the encoder network.

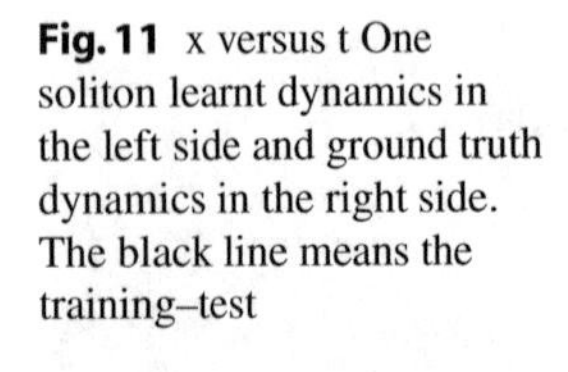

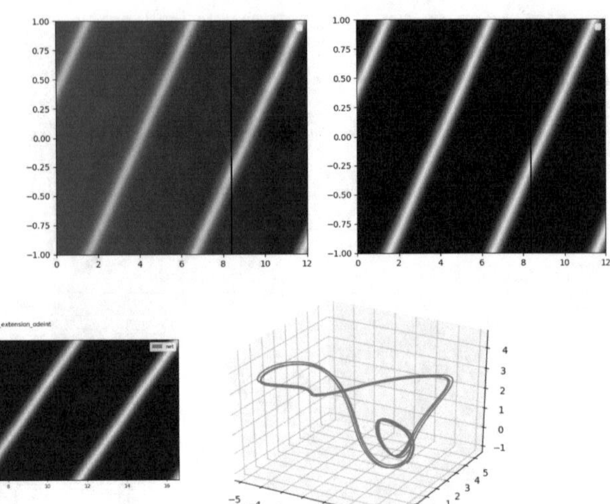

Fig. 11 x versus t One soliton learnt dynamics in the left side and ground truth dynamics in the right side. The black line means the training–test

Fig. 12 Time extension of the network integrated trajectory showing stable dynamics. Latent trajectory on the right side, corresponding to a conservative closed orbit, with a loop encoding the periodic boundary condition

Korteweg–de-Vries

Korteweg–de-Vries equation (23) was first conceived as model for waves on shallow water surfaces. It describes the behaviour of stable self-reinforcing wave packets called solitons, that arise from the equilibrium of dispersive and non-linear effects in the medium.

$$\frac{\partial u}{\partial t} + \frac{\partial^3 u}{\partial^3 x} - u\frac{\partial u}{\partial x} = 0. \tag{23}$$

This is one of the few PDEs with a known exact solution (24), corresponding to a functional family of soliton waves of different amplitude and speed.

$$\phi(x,t) = -\frac{1}{2} c \sec^2\left(\frac{\sqrt{c}}{2}(x - ct - a)\right). \tag{24}$$

In Fig. 11, learnt and ground truth dynamics are represented for one soliton.

In Fig. 12, the extended network integration of the learnt dynamics is shown in the X and h spaces, where the conservative character of the equation has been captured by the network.

As it was done for one soliton, in the case of several solitons, Figs. 13 and 14 are obtained. In Fig. 14, a two times training time latent trajectory is shown, proving as well the conservative character of the learnt solution, that appears in the latent space as a limit cycle. In Fig. 15 non seen test solutions, along with latent generated trajectories are displayed for the several solutions case.

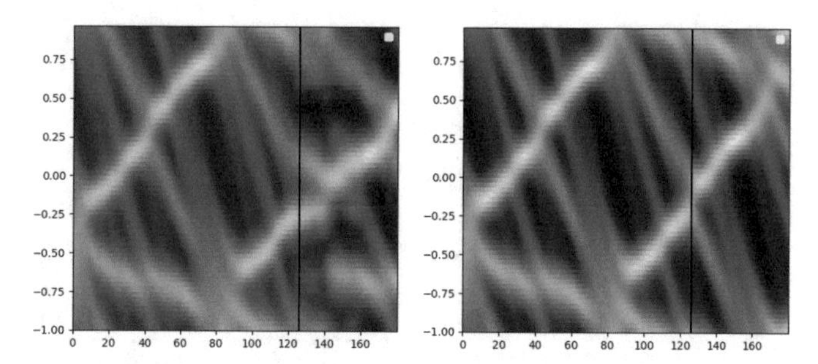

Fig. 13 x versus t Korteweg–de-Vries solution for an initial condition that splits mainly into three solitons. Learnt dynamics in the left side and real dynamics in the right side. The black line means the training–test

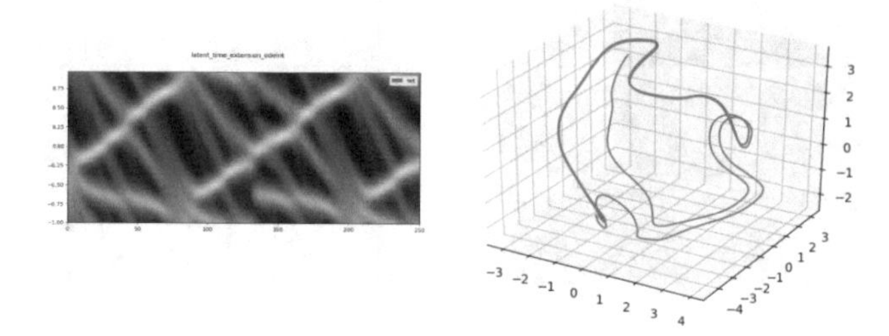

Fig. 14 Network time extension of the training trajectory, showing stable conservative dynamics. After the initial gaussian transient reaches the limit cycle

Kuramoto–Sivashinsky

Kuramoto–Sivashinsky equation (25) is a fourth-order non-linear PDE, that models laminar flame fronts with diffusive instabilities. The sign of the diffusive term acts as an energy source and it has a destabilizing effect. The convective uu_x term transfers energy to smaller wavelengths, where the fourth-order derivative term dominates and acts as a stabilizing factor, as it is explained with the dispersion relation $-iw = k^2 - k^4$. This richer set of mechanisms to transfer energy can lead to very complex behaviours, including chaoticity. Kuramoto–Sivashinsky is the simplest known PDE that exhibits this phenomenon.

$$\frac{\partial u}{\partial t} + \frac{\partial^4 u}{\partial^4 x} + \frac{\partial^2 u}{\partial^2 x} + u\frac{\partial u}{\partial x} = 0. \tag{25}$$

By using $L = 22$ as the domain width and the same initial condition solutions, KS has a positive Lyapunov exponent and exhibits chaotic behaviour. The intrinsic Hausdorff dimension of the chaotic attractor is greater than 3, so encoding it in a 3D space becomes complicated. In Fig. 16, it is shown the long-term prediction capability of a 4-neuron architecture. In Fig. 17 is shown as in chaotic dynamics, variations are expected for slightly different initial conditions.

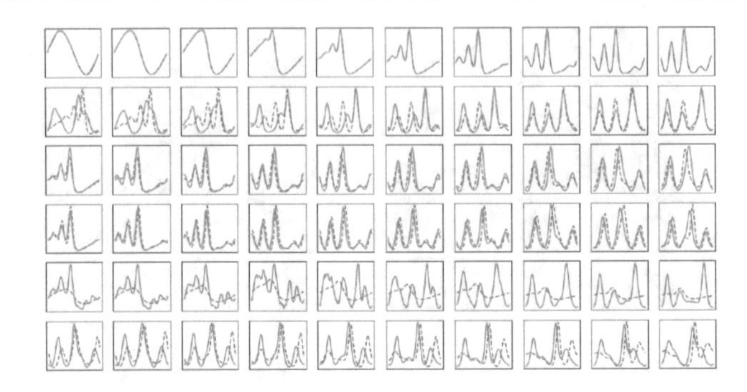

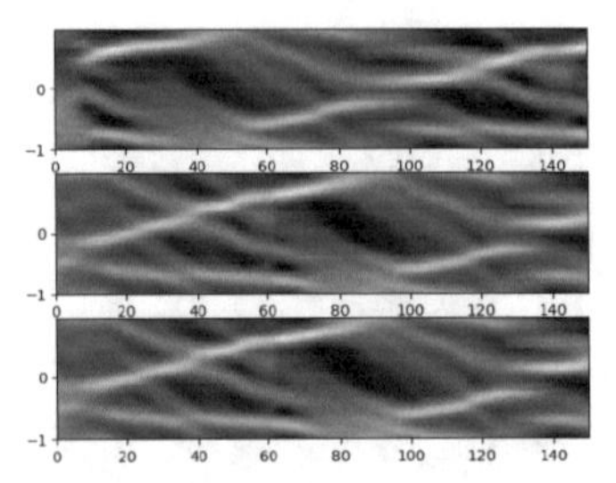

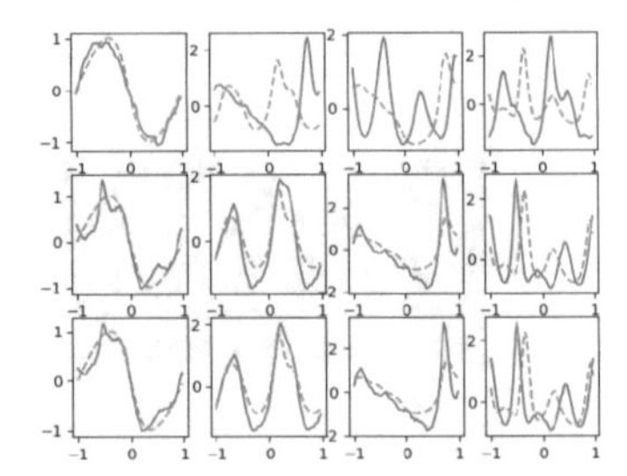

Fig. 15 In the upper side are represented non-seen test solutions, the first row belongs to the training trajectory. Down it is shown the mesh plot and time snapshots of different latent generated trajectories, moving in the first principal component axis of the latent activation space. In orange the training trajectory and the generated one in blue

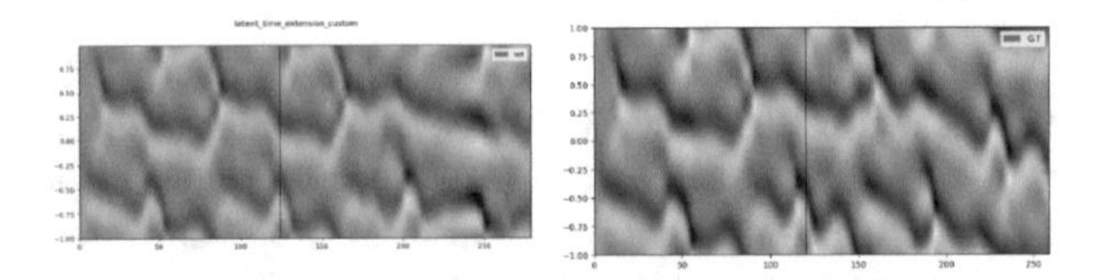

Fig. 16 On the left side,x versus t extended network dynamics. On the right side, real dynamics. The black line means the training-test

6 Conclusion

In this work, it was shown that a feed-forward properly regularized neural network, recast as a recurrent neural network and trained with the appropriate loss function, is able to qualitatively learn the global features of the phase space of dynamical systems, whose trajectories it was trained with. The examined characteristics were

1. Dissipative or conservative dynamics.
2. Limit cycles.
3. Unseen attractors.
4. Frequency response.
5. Parametric bifurcations.

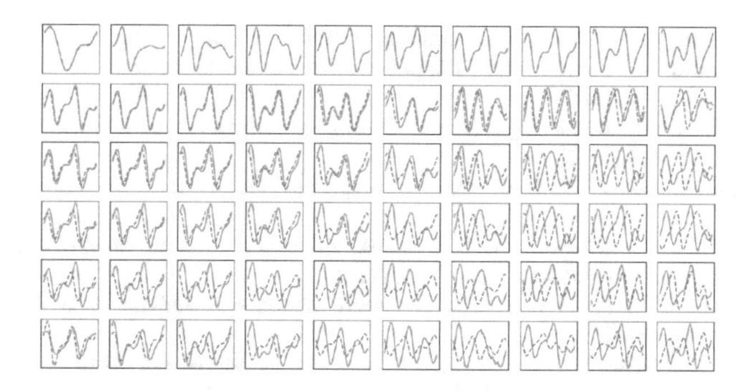

Fig. 17 Time snapshots of different test solutions. The first row corresponds to the training one. Neural network results in orange and ground truth in blue

It was found that even very small neural networks with non-linear $max(0, x)$ activations can exhibit complex behaviours. This fact allows them to qualitatively fit into the mould of very different dynamics, also when the difference between the original and learnt system derivative, in terms of MSE, is considerable. This promotes the possibility of using simple networks with adjustable precision to replicate the parts of a system's dynamics that interest us, with a much lower computational cost.

With the Lorentz system experiment, it became apparent how a neural network could faithfully learn a parameterized system, with little or no added capacity with respect to the single configuration system learning problem.

Advancing in the work, we have been able to encode non-linear 1D PDEs derivative into a very compact set of hidden variables. This has enabled us to solve a system of ODEs with much fewer degrees of freedom.

It was demonstrated that the encoded space was capable of solving trajectories with a variable degree of proximity to the learned one. As well as encoding complex abstract characteristics of motion such as advection velocities, dissipation rates or transient times.

Finally, in this reduced representation, the requirements for temporal discretization, that ensure stability, were significantly relaxed compared to the spectral solver used.

References

1. Kim, B., Azevedo, V.C., Thuerey, N., Kim, T., Gross, M., Solenthaler, B.: Deep fluids: a generative network for parameterized fluid simulations. Comput. Graph. Forum **38**, 59–70 (2019). Wiley Online Library
2. Wiewel, S., Becher, M., Thuerey, N.: Latent space physics: towards learning the temporal evolution of fluid flow. Comput. Graph. Forum **38**, 71–82 (2019). Wiley Online Library
3. Raissi, M., Perdikaris, P., Karniadakis, G.E.: Multistep neural networks for data-driven discovery of nonlinear dynamical systems (2018). arXiv:1801.01236
4. Funahashi, K.-I., Nakamura, Y.: Approximation of dynamical systems by continuous time recurrent neural networks. Neural Netw. **6**(6), 801–806 (1993)

5. Chen, T., Chen, H.: Universal approximation to nonlinear operators by neural networks with arbitrary activation functions and its application to dynamical systems. IEEE Trans. Neural Netw. **6**(4), 911–917 (1995)
6. Cybenko, G.: Approximation by superpositions of a sigmoidal function. Math. Control Signals Syst. **2**(4), 303–314 (1989)
7. Tsung, F.-S., Cottrell, G.W.: Phase-space learning. Adv. Neural Inform. Process. Syst., 481–488 (1995)
8. Trischler, A.P., D'Eleuterio, G.M.T.: Synthesis of recurrent neural networks for dynamical system simulation. Neural Netw. **80**, 67–78 (2016)
9. Navone, H.D., Ceccatto, H.A.: Learning chaotic dynamics by neural networks. Chaos Solitons Fractals **6**, 383–387 (1995)
10. Bakker, R., Schouten, J.C., Giles, C.L., Takens, F., van den Bleek, C.M.: Learning chaotic attractors by neural networks. Neural Comput. **12**(10), 2355–2383 (2000)
11. Pathak, J., Hunt, B., Girvan, M., Zhixin, L., Ott, E.: Model-free prediction of large spatiotemporally chaotic systems from data: A reservoir computing approach. Phys. Rev. Lett. **120**(2), 024102 (2018)
12. Pan, S., Duraisamy, K.: Long-time predictive modeling of nonlinear dynamical systems using neural networks. Complexity **2018** (2018)
13. Brunton, S.L., Proctor, J.L., Kutz, J.N.: Discovering governing equations from data by sparse identification of nonlinear dynamical systems. Proc. Natl. Acad. Sci. **113**(15), 3932–3937 (2016)
14. Wang, Z., Xiao, D., Fang, F., Govindan, R., Pain, C.C., Guo, Y.: Model identification of reduced order fluid dynamics systems using deep learning. Int. J. Numer. Methods Fluids **86**(4), 255–268 (2018)
15. Vlachas, P.R., Byeon, W., Wan, Z.Y., Sapsis, T.P., Koumoutsakos, P.: Data-driven forecasting of high-dimensional chaotic systems with long short-term memory networks. Proc. Royal Soc. A Math. Phys. Eng. Sci. **474**(2213), 20170844 (2018)
16. Rifai, S., Vincent, P., Muller, X., Glorot, X., Bengio, Y.: Contractive auto-encoders: explicit invariance during feature extraction. In: Proceedings of the 28th International Conference on International Conference on Machine Learning, pp. 833–840. Omnipress (2011)
17. Vincent, P., Larochelle, H., Bengio, Y., Manzagol, P.-A.: Extracting and composing robust features with denoising autoencoders. In: Proceedings of the 25th International Conference on Machine Learning, pp. 1096–1103. ACM (2008)
18. Kingma, D.P., Ba, J.: Adam: a method for stochastic optimization (2014). arXiv:1412.6980
19. Nair, V., Hinton, G.E.:. Rectified linear units improve restricted Boltzmann machines. In: ICML (2010)
20. He, K., Zhang, X., Ren, S., Sun, J.: Delving deep into rectifiers: surpassing human-level performance on imagenet classification. In: Proceedings of the IEEE International Conference on Computer Vision, pp. 1026–1034 (2015)
21. Osinga, H.M., Krauskopf, B.: Visualizing the structure of chaos in the Lorenz system. Comput. Graph. **26**(5), 815–823 (2002)
22. Montufar, G.F., Pascanu, R., Cho, K., Bengio, Y.: On the number of linear regions of deep neural networks. In: Advances in Neural Information Processing Systems, pp. 2924–2932 (2014)
23. Hirsch, M.W., Smale, S., Devaney, R.L.: Differential Equations, Dynamical Systems, and an Introduction to Chaos. Academic Press (2012)

Index

P. Quintela Estévez et al. (eds.), *Advances on Links Between Mathematics and Industry*, SxI - Springer for Innovation / SxI - Springer per l'Innovazione 15, https://doi.org/10.1007/978-3-030-59223-3

Printed in the United States
By Bookmasters